Praise for *On Being Certain*

"One of the startling implications of Burton's thesis is that we ultimately cannot trust ourselves when we believe we know something to be true. 'We can't afford to continue with the outdated claims of a perfectly rational unconscious or knowing when we can trust gut feelings,' he writes. *On Being Certain* challenges our understanding of the very nature of thought and provokes readers to ask what Burton calls 'the most basic of questions': How do we know what we know?"
—*Scientific American Mind*

"Neurologist Robert A. Burton, M.D., has written a gem of a book. The author is a neurologist who is also a novelist and a columnist for Salon.com. This well-written book is the result of many years of cogitation by a wise clinician. If there's anything you think you're certain of, read this book and you may change your mind."
—*Skeptical Inquirer*

"In his brilliant new book, Burton systematically and convincingly shows that certainty is a mental state, a feeling like anger or pride that can help guide us, but that doesn't dependably reflect objective truth. *On Being Certain* ought to be required reading for every citizen."
—*ForbesLife*

"What do we do when we recognize that a false certainty feels the same as certainty about the sky being blue? A lesser guide might get bogged down in nail-biting doubts about the limits of knowledge. Yet Burton not only makes clear the fascinating beauty of this tangled terrain, he also brings us out the other side with a clearer sense of how to navigate. It's a lovely piece of work; I'm all but certain you'll like it."
—David Dobbs, author of *Reef Madness: Charles Darwin, Alexander Agassiz, and the Meaning of Coral*

"Burton has a great talent for combining wit and insight in a way both palatable and profound."
—Johanna Shapiro, Ph.D., professor of Family Medicine at UC Irvine School of Medicine

"A new way of looking at knowledge that merits close reading by scientists and general readers alike." —*Kirkus Reviews*

"This could be one of the most important books of the year. With so much riding on 'certainty,' and so little known about how people actually reach a state of certainty about anything, some plain speaking from a knowledgeable neuroscientist is called for. If Gladwell's *Blink* was fascinating but largely anecdotal, Burton's book drills down to the real science behind snap judgments and other decision-making."
—Howard Rheingold, futurist and author of *Smart Mobs*

"A fascinating read. Burton's engaging prose takes us into the deepest corners of our subconscious, making us question our most solid contentions. Nobody who reads this book will walk away from it and say 'I know this for sure' ever again."
—Sylvia Pagán Westphal, science reporter, *The Wall Street Journal*

"Burton provides a compelling and thought-provoking case that we should be more skeptical about our beliefs. Along the way, he also provides a novel perspective on many lines of research that should be of interest to readers who are looking for a broad introduction to the cognitive sciences." —*Seed* magazine

"It's a wonderful book, easy to read, full of ideas, and its highlighting the 'feeling of certainty' as a topic of study is quite new, fertile, and genuinely unsettling."
—John Campbell, Willis S. and Marion Slusser Professor of Philosophy, UC Berkeley and author of *Reference and Consciousness*

On Being
CERTAIN

Also by Robert A. Burton, M.D.

Doc-in-a-Box
Cellmates

On Being
CERTAIN

BELIEVING YOU ARE RIGHT
EVEN WHEN YOU'RE NOT

Robert A. Burton, M.D.

 St. Martin's Griffin   New York

www.stmartins.com

Design by Elina D. Nudelman

The Library of Congress has catalogued the hardcover edition as follows:

Burton, Robert Alan, 1941–
   On being certain : believing you are right even when you're not / Robert A. Burton.—1st ed.
      p. cm.
   Includes bibliographical references and index.
   ISBN-13: 978-0-312-35920-1
   ISBN-10: 0-312-35920-9
  1. Certainty. I. Title.
   BD171 .B87 2008
   153.4—dc22

                                   2008001470

   ISBN-13: 978-0-312-54152-1 (pbk.)
   ISBN-10: 0-312-54152-X (pbk.)

10 9 8 7 6 5 4

for Adrianne

Contents

Preface

CERTAINTY IS EVERYWHERE. FUNDAMENTALISM IS IN FULL bloom. Legions of authorities cloaked in total conviction tell us why we should invade country X, ban *The Adventures of Huckleberry Finn* in schools, or eat stewed tomatoes; how much brain damage is necessary to justify a plea of diminished capacity; the precise moment when a sperm and an egg must be treated as a human being; and why the stock market will eventually revert to historical returns. A public change of mind is national news.

But why? Is this simply a matter of stubbornness, arrogance, and/or misguided thinking, or is the problem more deeply rooted in brain biology? Since my early days in neurology training, I have been puzzled by this most basic of cognitive problems: What does it mean to be convinced? At first glance, this question might sound foolish. You study the evidence, weigh the pros and cons, and make a decision. If the evidence is strong enough, you are convinced that there is no other reasonable answer. Your resulting sense of certainty feels like the only logical and justifiable conclusion to a conscious and deliberate line of reasoning.

But modern biology is pointing in a different direction. Consider for a moment an acutely delusional schizophrenic patient telling you with absolute certainty that three-legged Martians are secretly tapping his phone and monitoring his thoughts. The patient is utterly convinced of the "realness" of the Martians; he "knows" that they exist even if we can't see them. And he is surprised that we aren't convinced. Given what we now know about the biology of schizophrenia, we recognize that the patient's brain chemistry has gone amok, resulting in wildly implausible thoughts that can't be "talked away" with logic and contrary evidence. We accept that his false sense of conviction has arisen out of a disturbed neurochemistry.

It is through extreme examples of brain malfunction that neurologists painstakingly explore how the brain works under normal circumstances. For example, most readers will be familiar with the case of Phineas Gage, the Vermont laborer whose skull and frontal region of the brain were pierced with an iron bar during an 1848 railroad construction accident.[1] Miraculously, he lived, but with a dramatically altered personality. By gathering together information from family, friends, and employers, his physicians were able to piece together one of the earliest accurate descriptions of how the frontal lobe affects behavior.

Back to the pesky Martians. If Phineas Gage's posttraumatic personality change led to a better understanding of normal frontal lobe functions, is the schizophrenic's certainty that the Martians are listening to his thoughts a clue as to the origins of our sense of conviction? What is this patient telling us about the brain's ability to create the unshakable belief that what we feel that we know is, in fact, unequivocally correct? Are certainty and conviction purely deliberate, logical, and conscious choices, or not what they appear?

For me, the evidence is overwhelming; the answer is startling and counterintuitive, yet unavoidable. The revolutionary premise at the heart of this book is:

> Despite how certainty feels, it is neither a conscious choice nor even a thought process. Certainty and similar states of "knowing what we know" arise out of involuntary brain mechanisms that, like love or anger, function independently of reason.

To dispel the myth that we "know what we know" by conscious deliberation, the first section of the book will show how the brain creates the involuntary sensation of "knowing" and how this sensation is affected by everything from genetic predispositions to perceptual illusions common to all bodily sensations. Then we can see how this nonreasoned *feeling of knowing* is at the heart of many seemingly irresolvable modern dilemmas.

I am a neurologist with a novelist's sensibility. Though I have tried to make this book as accurate as possible, there will be many areas of controversy and frank disagreement. My goal is not to defend each argument against all criticism, but rather to generate a discussion about the nature and limitations of how we know what we know. To keep the book from being too dense or riddled with jargon, I have relegated more technical details, explanations, most personal digressions, and bibliography to the endnotes.

I must also confess to an underlying agenda: A stance of absolute certainty that precludes consideration of alternative opinions has always struck me as fundamentally wrong. But such accusations are meaningless without the backing of hard science. So I have set out to provide a scientific basis for challenging our belief in certainty. An unavoidable side effect: The scientific evidence will also show the limits of scientific inquiry. But in pointing out the biological limits of reason, including scientific thought, I'm not making the case that all ideas are equal or that scientific method is mere illusion. I do not wish to give ammunition to the legions of true believers who transform blind faith into evidence for creationism, alien abduction, or Aryan supremacy. The purpose is not to destroy the foundations of science, but only to point out the inherent limitations of the questions that science asks and the answers it provides.

My goal is to strip away the power of certainty by exposing its involuntary neurological roots. If science can shame us into questioning the nature of conviction, we might develop some degree of tolerance and an increased willingness to consider alternative ideas—from opposing religious or scientific views to contrary opinions at the dinner table.

A personal note: The schema that I am about to present has given me an unintended new way of seeing common problems. It isn't that I think about each issue and how it relates to neurobiology. Rather, the very notion of how we know—and even how we ask questions—has shaped how I feel and respond to everything from the daily news to pillow talk with my wife to age-old philosophical questions. The sense of an inner quiet born of acknowledging my limitations has been extraordinary; I would like to share this with you.

1

The Feeling of Knowing

I AM STUCK IN AN OBLIGATORY NEIGHBORHOOD COCKTAIL party during the first week of the U.S. invasion of Iraq. A middle-aged, pin-striped lawyer announces that he'd love to be in the front lines when the troops reach Baghdad. "Door-to-door fighting," he says, puffing up his chest. He says he's certain he could shoot an Iraqi soldier, although he's never been in a conflict bigger than a schoolyard brawl.

"I don't know," I say. "I'd have trouble shooting some young kid who was being forced to fight."

"Not me. We're down to dog-eat-dog."

He nods at his frowning wife, who's anti-invasion. "All's fair in love and war." Then back to me. "You're not one of those peacenik softies, are you?"

"It wouldn't bother you to kill someone?"

"Not a bit."

"You're sure?"

"Absolutely."

He's a neighbor and I can't escape. So I tell him one of my father's favorite self-mocking stories.

During the 1930s and '40s, my father had a pharmacy in one of the tougher areas of San Francisco. He kept a small revolver hidden beneath the back cash register. One night, a man approached, pulled out a knife, and demanded all the money in the register. My father reached under the counter, grabbed his gun, and aimed it at the robber.

"Drop it," the robber said, his knife at my father's throat. "You're not going to shoot me, but I *will* kill you."

For a moment it was a Hollywood standoff, *mano a mano*. Then my father put down his gun, emptied out the register, and handed over the money.

"What's your point?" the lawyer asks. "Your father should have shot him."

"Just the obvious," I say. "You don't always know what you're going to do until you're in the moment."

"Sure you do. I know with absolute certainty that I'd shoot anyone who was threatening me."

"No chance of any hesitation?"

"None at all. I know myself. I know what I would do. End of discussion."

MY MIND REELS with seemingly impossible questions. What kind of knowledge is "I know myself and what I would do"? Is it a conscious decision based upon deep self-contemplation or is it a "gut feeling"? But what is a gut feeling—an unconscious decision, a mood or emotion, an ill-defined but clearly recognizable mental state, or a combination of all these ingredients? If we are to understand how we know what we know, we first need some ground rules, including a general classification of mental states that create our sense of knowledge about our knowledge.

For simplicity, I have chosen to lump together the closely allied feelings of certainty, rightness, conviction, and correctness under the all-inclusive term, the *feeling of knowing*. Whether or not these are separate sensations or merely shades or degrees of a common feeling isn't important. What they do share is a common quality: Each is a form of metaknowledge—knowledge about our knowledge—that qualifies or colors our thoughts, imbuing them with a sense of rightness or wrongness. When focusing on the phenomenology (how these sensations *feel*), I've chosen to use the term the *feeling of knowing* (in italics). However, when talking about the underlying science, I'll use *knowing* (in italics). Later I will expand this category to include feelings of familiarity and realness—qualities that enhance our sense of correctness.

EVERYONE IS FAMILIAR with the most commonly recognized *feeling of knowing*. When asked a question, you feel strongly that you know an answer that you cannot immediately recall. Psychologists refer to this hard-to-describe but easily recognizable feeling as a tip-of-the-tongue sensation. The frequent accompanying comment as you scan your mental Rolodex for the forgotten name or phone number: "I know it, but I just can't think of it." In this example, you are aware of knowing something, without knowing what this sense of knowing refers to.

Anyone who's been frustrated with a difficult math problem has appreciated the delicious moment of relief when an incomprehensible equation suddenly *makes sense*. We "see the light." This *aha* is a notification from a subterranean portion of our mind, an involuntary all-clear signal that we have grasped the heart of a problem. It isn't just that we can solve the problem; we also "know" that we understand it.

Most *feelings of knowing* are far less dramatic. We don't ordinarily sense them as spontaneous emotions or moods like love or happiness; rather they feel like thoughts—elements of a correct line of reasoning. We learn to add $2+2$. Our teacher tells us that 4 is the correct answer. Yes, we hear a portion of our mind say. Something within us tells us that we "know" that our answer is correct. At this simplest level of understanding, there are two components to our understanding—the knowledge that $2+2=4$, and the judgment or assessment of this understanding. We know that our understanding that $2+2=4$ is itself correct.

The *feeling of knowing* is also commonly recognized by its absence. Most of us are all too familiar with the frustration of being able to operate a computer without having any "sense" of how the computer really works. Or learning physics despite having no "feeling" for the rightness of what you've learned. I can fix a frayed electrical cord, yet am puzzled by the very essence of electricity. I can pick up iron filings with a magnet without having the slightest sense of what magnetism "is."

At a deeper level, most of us have agonized over those sickening "crises of faith" when firmly held personal beliefs are suddenly stripped of a visceral sense of correctness, rightness, or meaning. Our most considered beliefs suddenly don't "feel right." Similarly, most of us have been shocked to hear that a close friend or relative has died unexpectedly, and yet we "feel" that he is still alive. Such upsetting news often takes time to "sink in." This disbelief associated with hearing about a death is an example of the sometimes complete disassociation between intellectual and felt knowledge.

To begin our discussion of the *feeling of knowing*, read the following excerpt at normal speed. Don't skim, give up halfway through, or skip to the explanation. Because this experience can't be duplicated once you know the explanation, take a moment to ask yourself how you feel about the paragraph. After reading the clarifying word, reread the paragraph. As you do so, please pay close attention to the shifts in your mental state and your feeling about the paragraph.

> A newspaper is better than a magazine. A seashore is a better place than the street. At first it is better to run than to walk. You may have to try several times. It takes some skill, but it is easy to learn. Even young children can enjoy it. Once successful, complications are minimal. Birds seldom get too close. Rain, however, soaks in very fast. Too many people doing the same thing can also cause problems. One needs lots of room. If there are no complications, it can be very peaceful. A rock will serve as an anchor. If things break loose from it, however, you will not get a second chance.

Is this paragraph comprehensible or meaningless? Feel your mind sort through potential explanations. Now watch what happens with the presentation of a single word: kite. As you reread the paragraph, feel the prior discomfort of something amiss shifting to a pleasing sense of rightness. Everything fits; every sentence works and has meaning. Reread the paragraph again; it is impossible to regain the sense of not understanding. In an instant, without due conscious deliberation, the paragraph has been irreversibly infused with a *feeling of knowing.*

Try to imagine other interpretations for the paragraph. Suppose

I tell you that this is a collaborative poem written by a third-grade class, or a collage of strung-together fortune cookie quotes. Your mind balks. The presence of this *feeling of knowing* makes contemplating alternatives physically difficult.

Each of us probably read the paragraph somewhat differently, but certain features seem universal. After seeing the word *kite*, we quickly go back and reread the paragraph, testing the sentences against this new piece of information. At some point, we are convinced. But when and how?

The kite paragraph raises several questions central to our understanding of how we "know" something. Though each will be discussed at greater length in subsequent chapters, here's a sneak preview.

- Did you consciously "decide" that *kite* was the correct explanation for the paragraph, or did this decision occur involuntarily, outside of conscious awareness?

- What brain mechanism(s) created the shift from not knowing to *knowing*?

- When did this shift take place? (Did you know that the explanation was correct before, during, or after you reread the paragraph?)

- After rereading the paragraph, are you able to consciously separate out the *feeling of knowing* that *kite* is the correct answer from a reasoned understanding that the answer is correct?

- Are you sure that *kite* is the correct answer? If so, how do you know?

2

How Do We Know
What We Know?

PARENTS' AND TEACHERS' CUSTOMARY ADVICE FOR "NOT getting" math and physics is to study harder and think more deeply about the problem. Their assumption is that more effort will bridge the gap between dry knowledge and felt understanding. Without this assumption, we would give up every time we failed to understand something at first glance. But for those "what's the point of it all" existential moments—when formerly satisfactory feelings of purpose and meaning no longer "feel right"—history and experience have taught us differently. Logic and reason rarely are "convincing." (In this context, "convincing" is synonymous with reviving this missing *feeling of knowing* what life is about.") Instead, we conjure up images of ascetics, mystics, and spiritual seekers—those who have donned hair shirts, trekked through the desert à la St. Jerome, huddled in caves or under trees, or sought isolation and silence in monasteries. Eastern religions emphasize a "stillness of the mind" rather than actively thinking about the missing sense of meaning.

So, which is it? Should the remedy for the absence of the *feeling of knowing* be more conscious effort and hard thought, or less? Or are both of these common teachings at odds with more basic neurobiology? Consider the curious phenomenon of *blindsight*, perhaps the best-studied example of the lack of the *feeling of knowing* in the presence of a state of knowledge.

Out of Sight Is Not Out of Mind

A patient has a stroke that selectively destroys his occipital cortex—the portion of the brain that receives primary visual inputs. His retina still records incoming information, but his malfunctioning visual cortex does not register the images sent from the retina. The result is that the patient consciously sees nothing. Now flash a light in various quadrants of his visual field. The patient reports that he sees nothing, yet he can fairly accurately localize the flashing light to the appropriate quadrant. He feels that he is guessing and is unaware that he is performing any better than by chance.

How is this possible?

First, let's trace the pathway of the "unseen" light. Some fibers from the retina proceed directly to the primary visual cortex in the occipital lobe. But other fibers bypass the region responsible for conscious "seeing" and instead project to subcortical and upper brain stem regions that do not produce a visual image. These lower brain areas are primarily concerned with automatic, reflexive functions such as fight-or-flight. Quickly approaching or looming objects cause the body to swing the head into position so that the eyes can examine the threat. An immediate reflexive action has clear evolutionary benefits over more time-consuming conscious perception and deliberation. In the broadest sense, you could say that these subcortical regions "see" the threat without sending a visual image into awareness.

Blindsight is a primitive unconscious visual localization and navigation system uncovered by the patient's cortical blindness. The patient's subliminal knowledge of the location of the flashing light doesn't trigger the *feeling of knowing* because news of this knowledge can't reach the higher cortical regions that generate the feeling. As a result, the patient swears that he hasn't seen a flashing light, yet he clearly possesses a subliminal knowledge of the light's location. When he chooses the proper visual field for the flashing light, he has no feeling that this is a correct answer. *He does not know what he knows.*[1]

With blindsight, we see the disconnect between knowledge and awareness of this knowledge as being related to a fundamental flaw in our circuitry. This broken connection cannot be restored either through conscious effort or stilling of the mind—the problem is not within our control.

Though clinically apparent blindsight is a rare event usually caused by a stroke that interferes with the blood supply to the occipital cortex, faulty expressions of the *feeling of knowing* are everyday occurrences. Let's begin with our own memories.

The *Challenger* Study

Try to remember where you were when Kennedy was assassinated, the *Challenger* blew up, or the World Trade Center was attacked. Now ask yourself how certain you are of those memories. If you believe that you are quite sure of where you were when you heard the news, keep that feeling in mind as you read about the *Challenger* study in the next pages. If you don't remember where you were, ask yourself how you know that you don't remember. (Keep in mind the blindsight example when asking this question.) Either way, try to understand the feeling and your degree of certainty of this memory.

At my most recent med school reunion dinner, several former classmates were recalling where they were when Kennedy was assassinated. We had been in the second year of medical school, which meant that we all went to the same classes. Wherever one was, we probably all were. But the recollections were strikingly different; after dinner the discussion was becoming increasingly heated, as though each classmate's mind was on trial. A urologist thought we were at lunch, an internist said we were in the lab. A pathologist remembered being at a pub down the street from the med center. "That can't be true," the urologist said. "The assassination was at noon, Dallas time. You didn't go to the bars 'til after class."

I laughed and briefly described the *Challenger* study.[2]

Within one day of the space shuttle *Challenger* explosion, Ulric Neisser, a psychologist studying "flashbulb" memories (the recall of highly dramatic events), asked his class of 106 students to write down exactly how they'd heard about the explosion, where they were, what they'd been doing, and how they felt. Two and a half years later they were again interviewed. Twenty-five percent of the students' subsequent accounts were strikingly different than their original journal entries. More than half the people had lesser degrees of error, and less than ten percent had all the details correct. (Prior to seeing their original journals, most students presumed that their memories were correct.)

Most of us reluctantly admit that memory changes over time.

As kids, we saw how a story changed with retellings around a campfire. We have been at enough family reunions to hear once-familiar shared events morphed into unrecognizable and often contradictory descriptions. So, seeing that your journal entries were different than your recollection a couple of years later shouldn't be surprising. What startled me about the *Challenger* study were the students' responses when confronted with their conflicting accounts. Many expressed a high level of confidence that their false recollections were correct, despite being confronted with their own handwritten journals. The most unnerving was one student's comment, "That's my handwriting, but that's not what happened."

Why wouldn't the students consider their journal entries written shortly after the event to be more accurate than a recollection pulled up several years later? Pride, stubbornness, or fear of admitting an error? Not remembering the details of the *Challenger* explosion doesn't imply some massive personal failing that would make resistance to contrary evidence so overwhelming. Conversely, wouldn't pride in being logical and rational steer the students toward choosing their own handwriting over memories that they know might have been altered with time?

The inflamed urologist interrupted me, insisting the pathologist concede that he was wrong. The pathologist refused, turned to me, and said, "You tell them, Burton. You were there in the bar with me."

"Beats me. I just don't remember."

"That's not possible," the two warring doctors said simultaneously. "Everyone remembers the Kennedy assassination."

I shrugged and silently marveled at the vehemence of my classmates' convictions. Even telling them of the *Challenger* study persuaded no one, as though they were intent upon reproducing the very study that questioned their recollections. All felt that they were right, that they absolutely *knew* where they were and what they were doing when Kennedy was assassinated.

Cognitive Dissonance

In 1957, Stanford professor of social psychology Leon Festinger introduced the term *cognitive dissonance* to describe the distressing mental state in which people "find themselves doing things that don't fit with what they know, or having opinions that do not fit with other opinions they hold."[3] In a series of clever experiments, Festinger demonstrated that such tensions were more often minimized or resolved through changes in personal attitudes than by relinquishing the dissonant belief or opinion.

As an example, Festinger and his associates described a cult that believed that the earth was going to be destroyed by a flood. When the flood did not happen, those less involved with the cult were more inclined to recognize that they had been wrong. The more invested members who had given up their homes and jobs to work for the cult were more likely to reinterpret the evidence to show that they were right all along, but that the earth was not destroyed because of their faithfulness.[4]

Festinger's seminal observation: The more committed we are to a belief, the harder it is to relinquish, even in the face of overwhelming contradictory evidence. Instead of acknowledging an error in judgment and abandoning the opinion, we tend to develop a new attitude or belief that will justify retaining it. By giving us a model to consider how we deal with conflicting values, the theory of cognitive dissonance has become one of the most influential theories in social psychology. Yet it fails to convincingly answer why it is so difficult to relinquish unreasonable opinions, especially in light of seemingly convincing contrary evidence. It is easy to dismiss such behavior in cult members and others "on the fringe," but what about those of us who presume ourselves to be less flaky, those of us who pride ourselves on being levelheaded and reasonable?

WE MIGHT THINK of the *Challenger* study as an oddity, but here are additional examples of consciously choosing a false belief because it *feels* correct even when we know better. I have chosen the first example as a prelude to a later discussion in chapter 13 of the deeply rooted biological component of the science-versus-religion struggle. The second example, highlighting the cognitive dissonance of the placebo effect, introduces the idea that an unjustified *feeling of knowing* can have a clear adaptive benefit.

A Scientist Contemplates Creationism

Kurt Wise, with a B.A. in geophysics from the University of Chicago, a Ph.D. in geology from Harvard, where he studied under Steven Jay Gould, and a professorship at Bryan College in Dayton, Tennessee, writes of his personal conflict between science and religion.[5]

> I had to make a decision between evolution and Scripture. Either the Scripture was true and evolution was wrong or evolution was true and I must toss out the Bible. . . . It was there that night that I accepted the Word of God and rejected all that would ever

counter it, including evolution. With that, in great sorrow, I tossed into the fire all my dreams and hopes in science. . . . *If all the evidence in the universe turns against creationism, I would be the first to admit it, but I would still be a creationist because that is what the Word of God seems to indicate.* (Italics mine.)

A Patient Confronts the Placebo Effect

In a study of 180 people with osteoarthritis of the knee, a team of Houston surgeons headed by Bruce Moseley, M.D., found that patients who had "sham" arthroscopic surgery reported as much pain relief and improved mobility as patients who actually underwent the procedure.[6]

Mr. A, a seventy-six-year-old retired World War II veteran with a five-year history of disabling knee pain from X-ray-documented degenerative osteoarthritis was assigned to the placebo group (sham surgery in which general anesthesia was given, superficial incisions were made in the skin over the knee, but no actual surgical repair was performed). After the procedure, Mr. A was informed that he had received sham surgery; the procedure was described in detail. Nevertheless, he dramatically improved; for the first time in years he was able to walk without a cane. When questioned, he both fully understood what sham surgery meant and fully believed that his knee had been fixed.

"The surgery was two years ago and the knee has never bothered me since. It's just like my other knee now. I give a whole lot of credit to Dr. Moseley. Whenever I see him on the TV, I call the wife in and say, 'Hey, there's the doctor that fixed my knee!' "[7]

Our creationist geologist cringes at his own irrationality and yet declares that he does not have a choice. A patient "knows" that he hasn't had any reparative surgery performed, yet insists that the doctor fixed his knee. What if we could find patients who developed similar difficulties with reason as the result of specific brain insults (lesions)? If brain malfunctions can produce a similar flawed logic, what might that tell us about the biological underpinnings of cognitive dissonances?

Cotard's Syndrome

Ms. B, a twenty-nine-year-old grad student hospitalized for an acute viral encephalitis (a viral inflammation of the brain) complained: "Nothing feels real. I am dead." The patient refused any medical care. "There is no point in treating a dead person," she insisted. Her internist tried to reason with her. He asked her to put her hand on her chest and feel her heart beating. She did, and agreed that her heart was beating. He suggested that the presence of a pulse must mean that she was not dead. The patient countered that, since she was dead, her beating heart could not be evidence for being alive. She said she recognized that there was a logical inconsistency between being dead and being able to feel her beating heart, but that being dead felt more "real" than any contrary evidence that she was alive.

Weeks later, Ms. B began to recover; eventually she no longer believed that she was dead. She was able to make a distinction between her recovered "reality" and her prior delusions, yet she continued to believe that it must be possible to feel one's heart beat after death. After all, it had happened to her.

Cotard's syndrome—*le délire de négation*—is attributed to a French psychiatrist, Jules Cotard, who in 1882 described several patients with delusions of self-negation. These ranged from the belief that parts of the body were missing, or had putrefied, to the complete denial of bodily existence. The syndrome has been described with a variety of brain injuries, strokes, and dementia, as well as severe psychiatric disorders. The most extraordinary element of the syndrome is the patient's unshakable belief in being dead that overpowers any logical counterconclusion. Feeling one's beating heart isn't sufficient evidence to overcome the more powerful sense of the reality of being dead.

Other delusional syndromes associated with acute brain lesions include believing that a friend or a relative has been replaced by an impostor, or a double, or has taken on different appearances or identities, or that an inanimate object has been replaced by an inferior copy. The clinical feature common to all of these syndromes is the inability of the patient to shake a belief that he logically knows is wrong.

Mr. C, an elegant retired art dealer, was hospitalized overnight with a small stroke. The next morning, he felt well and was discharged. Within moments of returning home, he phoned my office in a panic. He was certain that his favorite antique desk had been replaced by a cheap Levitz reproduction. "Hurry over and see for yourself." He lived near my office; I dropped by at lunchtime. The desk in question was a massive eighteenth-century Italian refectory table that took up most of his den. It could easily seat a dozen; just lifting it would require a minimum of several men. And it was far too wide to fit through the doorway without removing the French doors. I quickly pointed out the impossibility of someone sneaking in, moving out the desk, and substituting a fake. Mr. C shook his head. "Yes, I admit that it is physically impossible that the desk has been replaced. But it has. You have to take my word for it. I know real when I see real, and this desk isn't real." He ran his hand along the grain, repeatedly fingering a couple of prominent wormholes. "It's funny," he said with a puzzled expression. "These are exact replicas of the holes in my desk. But they don't feel the least bit familiar. No," he announced emphatically, "someone must have replaced it." He then delivered the cognitive checkmate: "After all, I know what I know."

Although not restricted to a single area of the brain or a single definitive physiology, the most striking shared characteristic of these *delusional misidentification syndromes* is that the conflict between logic and a contrary *feeling of knowing* tends to be resolved in favor of feeling. Rather than rejecting ideas and beliefs that defy common sense and overwhelming contrary evidence, such patients end up using tortured logic to justify the more powerful sense of *knowing what they know*.[8]

Mr. C's statements also point out that *knowing* may also involve additional hard-to-define mental states such as a sense of familiarity and feelings of "realness." Like the tip-of-the-tongue sensation or the feeling of déjà vu, a sense of being familiar suggests some prior experience or knowledge. When stumped on a multiple choice test question, we tend to choose the answer that feels most familiar. Though we have no justification, we presume that such answers are more likely to be correct than those that we don't recognize or seem unfamiliar. Mr. C's "I know real when I see real" points out how a sense of "realness" might also bias us toward believing that an idea is correct. Patients with delusional misidentification syndromes often use "correct" interchangeably with "real."

It is likely that Mr. C's stroke affected his ability to appropriately experience feelings of familiarity and "realness." When neither the sight nor the feel of the desk triggered these feelings, he was forced to conclude that this desk could not be the original. Such delusions might be seen as an attempt to resolve a cognitive dissonance between hard evidence (the table is too big to move) and the absence of any feeling of familiarity and realness when Mr. C examined his desk.

In chapter 3 we shall see that the mental states of familiarity, "realness," conviction, truth, déjà vu, and tip of the tongue share a similar physiology with the *feeling of knowing*, including the ability to be directly triggered with electrical stimulation of the brain's limbic system.

It May Be Right, But It's Not *Right*

The other day, at a downtown garage, I left my car with valet parking. I returned, started to drive away, but felt something was wrong. I questioned the attendant's gaze, wondering if I'd paid too much. I checked the gas and oil gauge, and whether one of the doors was ajar. Then I realized that the seat had been readjusted by the attendant. It was a nominal difference, the seat was at most a half inch higher than usual. My derriere knew immediately; it took me considerably longer.

I was reminded of a story attributed to Ludwig Wittgenstein.

A man walks into a tailor's shop. The sign over the front door reads: CUSTOMER SATISFACTION GUARANTEED. The man orders a custom-made suit that should fit exactly like the one he is wearing. The tailor painstakingly measures every detail and jots them down in a notebook. A week later the customer returns to try on the new suit.

"It's not right," the customer says with annoyance. . . .

"Of course it is," the tailor says. "Here, I'll show you." The tailor takes out his measuring tape, compares the suit's readings with those in his notebook. "See, they're identical."

The customer shifts in his new suit but is still uncomfortable and displeased. "It may be right, but it's not *right*." He refuses to pay for the suit and storms out.

In the case of my car seat, I was forced to think through all the possible reasons that I sensed something was wrong. Fortunately, there was something measurable (the new angle of the car seat) that *explained* what I was *feeling*. With the tailor example, the customer's sense of something amiss is a matter of taste, of inexpressible or subconscious aesthetics. No matter what the measurements, the suit does not *feel* right.

The tailor demands his money; the customer admits that the suit was to his specifications, but not to his liking, and therefore he is under no obligation to buy the suit. Each *feels* that he is right. Hence that irritating popular refrain—end of discussion. We often talk about gut feelings. There is now extensive literature on the neuroenteric brain, as though some form of thought might actually originate in the pit of your stomach. Maybe so. And maybe my body just *knew* that my car seat was out of whack. But whatever the origin of the sensation, the key feature is that there seems to be an underlying *sense* or *feeling* that something is either correct or incorrect.

Consider the similarity in tone between the *Challenger* study student who said, "That's my writing, but that's not what happened," and the suit customer's "It might be right, but it's not *right*." When such a sense of conviction overrides obvious logical inconsistencies or scientific evidence, what is happening? Is it possible that there is an underlying neurophysiological basis for the specific sensation of *feeling right* or of *being right* that is so powerful that ordinary rational thought *feels* either wrong or irrelevant? Conviction versus knowledge—is the jury rigged, the game fixed by a basic physiology hidden beneath awareness?

3

Conviction Isn't a Choice

It is no great accomplishment to hear a voice in your head. The accomplishment is to make sure that it is telling you the truth.

—A patient describing a near-death experience

THE STUDIES OF BLINDSIGHT DEMONSTRATE THAT KNOWL-edge and the awareness of this knowledge arise from separate regions of the brain. So, we should also be able to find clinical examples of the converse of blindsight—moments of abnormal or altered brain function when the expression of the *feeling of knowing* occurs in the absence of any knowledge.

Of course, at first glance, the very idea of an isolated *feeling of knowing* seems ludicrous. A sense of knowledge, to have any meaning, must refer to something "known." We know "something," not "nothing." To dispel this notion that a *feeling of knowing* must be attached to a thought, this chapter will briefly touch on such seemingly unrelated phenomena as spontaneous and chemi-cally induced religious experiences, Dostoyevsky's epileptic aura, as well as detailed temporal lobe stimulation studies.

To experience the range of these states of *knowing* unassoci-ated with any specific knowledge, let's begin with the century-old classic—the *Varieties of Religious Experience* by William James—

which, for me, remains one of the most elegant testimonials to the power of clinical observations to explore the mind. James offers these illuminating quotes followed by his own comments (italics in these excerpts are mine).

Alfred Lord Tennyson:

I have never had any revelations through anesthetics, but a kind of waking trance—this for lack of a better word—I have frequently had, quite up from boyhood, when I have been all alone. This has come upon me through repeating my own name to myself silently, till all at once, as it were out of the intensity of the consciousness of individuality, individuality itself seemed to dissolve and fade away into boundless being, and *this not a confused state but the clearest, the surest of the surest, utterly beyond words.* . . . By God Almighty! There is no delusion in the matter! *It is no nebulous ecstasy, but a state of transcendent wonder, associated with absolute clearness of mind.*[1]

Saint Teresa:

One day, it was granted me to perceive in one instant how all things are seen and contained in God. I did not perceive them in their proper form, and nevertheless the view I had of them was of a sovereign clearness, and has remained vividly impressed upon my soul. . . . The view was so subtle and delicate that the understanding cannot grasp it.[2]

James's summary opinion:

Personal religious experience has its root and centre in mystical states of consciousness. . . . Its quality must be directly experi-

enced; it cannot be imparted or transferred to others. In this pe-
culiarity, mystical states are more like states of feeling than like
states of intellect. . . . *Although so similar to states of feeling, mystical states seem to those who experience them to be also states of
knowledge.* They are states of insight into depths of truth un-
plumbed by the discursive intellect. They are illuminations, revelations, full of significance and importance, all inarticulate though
they remain; and as a rule they carry with them a curious sense of
authority for after-time.[3]

This is a brilliant observation, equating religious and mystical
states with the sensation of *knowing,* and with the further recognition that such knowledge is felt, not thought. Though lacking in
modern-day neuroscience techniques, James was able to put his
finger directly on a key feature of how we know what we know:
"Mystical truth . . . resembles the knowledge given to us in sensations more than that given by conceptual thought."[4]

James's description is perfectly straightforward—with mystical states, people experience spontaneous mental sensations that
feel like knowledge but occur in the absence of any specific
knowledge. Felt knowledge. Knowledge without thought. Certainty without deliberation or even conscious awareness of having had a thought.

Neurotheology

In James's time, speculations on the cause of religious epiphanies
fell into two major camps: the psychological—hysteria, conversion
reaction, schizoid personality disorder, and so on—or the spiritual,
with claims of direct revelation from a higher power. Now we are
increasingly hearing of a third possibility. Recent neurophysiological

studies suggest that such feelings arise directly from the activation of localized areas of the brain (the limbic system)—either spontaneously or as the result of direct stimulation. According to UCLA neurologist Jeffrey Saver, this is the most compelling explanation for the mystical experiences of Saint Paul, Mohammad, Emanuel Swedenborg, Joseph Smith, Margery Kempe, Joan of Arc, and Saint Teresa.[5] The passage most commonly cited by neurologists is from a journal entry of Dostoyevsky. Though we lack pathological confirmation, the nature of Dostoyevsky's seizures is typical of seizures arising from disorders of the temporal lobe–limbic system structures.

On Easter eve night, circa 1870, Dostoyevsky is talking with a friend about the nature of God. Suddenly he cries out, "God exists; he exists." Then he loses consciousness, experiencing an epileptic fit. Dostoyevsky later wrote in his journal:

> I felt that heaven was going down upon the earth and that it had engulfed me. I have really touched God. He came into me myself, yes. You all, healthy people, can't imagine the happiness which we epileptics feel during the second before our fit. . . . I don't know if this felicity last for seconds, hours or months, but believe me, for all the joys that life may bring, I would not exchange this one.[6]

Ecstatic bliss triggered solely by wayward electrons? Why not? If you accept the studies of Toronto psychologist Michael Persinger, the same effect can be created with external stimulation of the brain. Volunteers don a cloth swimmer's cap outfitted with a grid of magnetic coils. Using the magnets to stimulate localized areas of the brain, Persinger has been able to generate feelings of a "sensed presence," "another self," or "oneness with the universe" (actual patient descriptions). Those with a Christian upbringing often describe the presence of Jesus; those with Muslim backgrounds have described the presence of Mohammad. Also frequently mentioned are profound emotions such as awe, joy, and a general sense of harmony and deep significance—though without being attached to any specific idea or belief.

It isn't surprising that there is an ever-growing literature on the biological origin of the religious impulse, for example, *Why God Won't Go Away* and *The "God" Part of the Brain*, or that my mailbox is stuffed with invites to weekend conferences on "neurotheology." The underlying point is both profound and self-evident: Even if the origination of the sense of God were extracorporeal—from a distant black hole, a past life, a dead relative, the rings around Uranus, or God in his or her heaven—the final pathway for the message's perception must reside within the brain.

Chemical activation of mystical states is as old as the most ancient psychedelic. William James described the phenomena with several anesthetics—chloroform, ether, and nitrous oxide. The following chloroform-induced mystical experience is a good example of a chemically induced cognitive dissonance: The knowledge that the mystical experience is a result of mundane chemistry does not negate the nagging (and lingering) sense of the certainty of God's existence. Note also that chloroform evoked the sensations of purity and truth *without any reference to any specific idea or thought*.

I cannot describe the ecstasy I felt. Then, as I gradually awoke from the influence of the anesthetics, the old sense of my relation to the world began to return, and the new sense of my relation to

God began to fade. . . . Think of it. To have felt purity and tenderness and truth and absolute love, and then to find that I had after all had no revelation, but that I had been tricked by the abnormal excitement of my brain. *Yet, this question remains. Is it possible that the inner sense of reality . . . was not a delusion, but an actual experience? Is it possible that I felt what some of the saints have said that they always felt, the undemonstrable but indisputable certainty of God?*[7] (Italics mine.)

In the following ether-induced example, another subject confirms the power of the mystical experience to feel as if a greater knowledge than objective evidence: "In that moment the whole of my life passed before me, including each little meaningless piece of distress, and I understood them. This was what it had all meant, this was the piece of work it had all been contributing to do. . . . *I perceived also in a way never to be forgotten, the excess of what we see over what we can demonstrate.*"[8] (Italics mine.)

Volunteers undergoing intravenous ketamine infusions (an anesthetic molecularly similar to the street drug PCP or angel dust), frequently experience a profound clarity of thought. One subject described "a sense of understanding everything, of knowing how the universe works."[9] Such descriptions are quite similar to those who've had "near-death experiences" from a cardiac arrest or an anesthetic complication; indeed, there may be a common mechanism of action.[10] Lack of adequate brain oxygen characteristically triggers the release of the neurotransmitter glutamate. Under normal conditions glutamate binds to NMDA receptors; in excessive amounts it is neurotoxic and facilitates neuronal death. In an attempt to prevent this cell death, the oxygen-deprived brain also releases protective chemicals that block the effect of glutamate on NMDA receptors. Ketamine has a similar NMDA receptor-blocking effect. So does MDMA (Ecstasy), another psychoactive drug known to produce feelings of mental clarity.[11] It is now believed that this blocking of the NMDA receptor is responsible for the clinical picture of a near-death experience.

Voices from the Limbic System

With each of the earlier descriptions, we are at the mercy of brief, highly emotionally charged, and difficult to reproduce patient reactions. Fortunately, we have a more consistent, controlled, and reproducible method for eliciting these mental states of *knowing*— formal brain stimulation/mapping of the temporal lobe–limbic system. As we proceed, keep in mind that brain mapping is the same technique that neurologists have used to localize other primary brain functions such as motor movements, vision, and hearing. But first, a word about the limbic system.

Though some neuroscientists question its existence as a specific entity,[12] the term *limbic system* is useful for discussing those regions of the brain fundamental to the most primary and basic emotions.[13] It includes the evolutionarily oldest regions of the cortex and subcortex—the cingulate gyrus, amygdala, hippocampus, the hypothalamus, and a variety of basal forebrain structures including the ventral tegmental area (the site of the brain's primary reward system), as well as associated regions of the frontal cortex that are implicated in emotional responses and decision making.[14]

Unfortunately for lab animals, the easiest emotion to study is good old-fashioned terror. Enter Joseph LeDoux, professor of neuroscience at New York University, with his provocative and

ingenious series of experiments. LeDoux conditioned rats to as-
sociate the sound of a ringing bell with electric shocks applied
to their paws. After being conditioned, the sound of the bell,
without the electric shocks, was sufficient to provoke a typical
fear response—momentary cessation of body movement, change
in heart rate, blood pressure, sweating, and release of stress hor-
mones.[15] LeDoux set out to find out the pathways that produced
this fear response.

He found that cutting the rats' acoustic nerves—the neural connection between the ears and the brain—abolished the fear response. (The sound of the bell didn't reach the brain.) If he left
the nerves intact, but surgically removed the auditory cortex—the
region of brain that processes and creates the conscious awareness
of sounds—the rats no longer "heard" the sound, yet the fearful
behavior persisted.[16] Just as the phenomenon of blindsight is
based upon visual images being transmitted to and processed in areas other than the visual cortex, LeDoux surmised that the sound
of the bell reached areas of subcortical brain capable of triggering
the fear response without the rat consciously hearing the bell.
LeDoux was able to demonstrate the presence of neural pathways
that bypass the auditory cortex, connecting directly with a temporal lobe structure—the amygdala—long known to be crucial to the
recognizing, processing, and remembering of emotional reactions,
including the fear response. From the amygdala these nerve fiber
pathways continue to regions of the hypothalamus that control
the sympathetic nervous system leading to increased heart rate,
blood pressure, and sweating, as well as to regions of the brain
stem that control reflexes and the facial expressions of fear.

LeDoux's experiments greatly clarified the role of the amygdala
in evoking a fear response without the need for any conscious

awareness and recognition of the provoking stimulus.[17] Other experiments have confirmed that direct stimulation of the amygdala produces the same fear response as Ledoux's conditioning experiments. Conversely, bilateral removal of the amygdala in animals, from rats to monkeys, produces a state of utter fearlessness. Knocking out a single gene active in the amygdala can greatly diminish the fear response in rats.

This fearlessness has also been observed in those rare patients with bilateral amygdala damage. Such patients characteristically approach new and potential risky situations with a positive, unafraid attitude. One man with bilateral amygdala damage loved to hunt deer in Siberia while dangling from a helicopter. Another extensively studied patient, SM, a young woman with calcification and atrophy of both amygdalae, could not be startled by the unexpected blast of a 100 decibel boat horn. Despite repeated conditioning attempts, SM did not demonstrate any autonomic changes—such as rise in pulse or blood pressure.[18] According to Antonio Damasio, the behavioral neurologist who has extensively investigated her deficits, SM can intellectually discuss what fear is, but the bilateral damage to her amygdala has prevented her from learning the significance of potentially dangerous situations.[19] (In chapter 9, we will return to the amygdala's role in processing and creating memories of fearful events.)

As the result of such studies, neurologists now accept that the amygdala is necessary for the expression of fear. But the study of mental states that defy precise classification—such as déjà vu or a sense of dread—is much more difficult. We have problems both in what to call them and how to standardize our observations. It is easy to recognize a scared rat, but a rodent's sense of alienation is less obvious. As a consequence, there are few formal and systematic studies; the closest that we have are the informal investigations carried out during the evaluation of patients with a particular form of epilepsy that originates from temporal lobe–limbic structures.

Most commonly as the result of a birth injury and developmental abnormalities, and occasionally due to a tumor, a patient can develop a particular form of epilepsy—a complex partial seizure. These spontaneous electrical discharges from temporal lobe–limbic structures characteristically produce a transient (seconds to minutes) alteration or clouding of consciousness, often associated with the intrusion of other mental feelings—déjà vu, dread, fear, and even religious feelings such as those described by Dostoyevsky. Their intensity varies from brief lapses in awareness to a complete loss of consciousness and major convulsions. The frequency also varies greatly. Some patients have very few seizures that are completely controlled with medication; others less fortunate can experience upwards of several dozen seizures per day despite maximal medication.

For the latter group, surgical removal of the damaged area of temporal lobe can result in a striking reduction or a cessation of seizures. As the major risk of surgery is creating damage to adjacent vital areas, the operating neurosurgeon must first identify the functions of all surrounding brain tissue. The surgery can be performed under local anesthesia (the brain is insensitive to pain); patients remain conscious and are able to describe exactly what they are experiencing. The surgeon systematically stimulates small areas of cerebral cortex; patient responses are recorded. At the conclusion of this cortical mapping, the surgeon has an excellent correlation between brain anatomy and its function and can better avoid operating near critical areas.

For our discussion, I've chosen three detailed series of operative brain mapping—temporal lobe stimulations that provide the most in-depth patient descriptions. To avoid possible cultural bias, I've included studies of patients from Canada, France, and Japan.[20] Despite the obvious differences in background, culture, and language, the similarities remain striking. Though I've grouped patient descriptions according to general categories of experience, there is some degree of obvious overlap. Also, many of these "feelings" occur either concomitantly or in rapid succession. I have also included some descriptions of the patient's spontaneous seizures. Cortical stimulations are labeled CS; spontaneous seizures are labeled SZ. Each description is from a different patient. All italics are mine.

As you listen to these voices of the limbic system, keep in mind that what these patients describe is not dependent upon any specific antecedent thought, line of reasoning, mood, personality quirk, or circumstance. A jolt of electricity is all that is necessary.

Déjà Vu and Feelings of Familiarity

SZ: "I do not know where it is, but *it seems very familiar* to me. . . . I feel very close to an attack—I think I am going to have one—a familiar memory."

CS: "I have the impression of already having been here, that I had already lived through this."

SZ: "Patient stated that a thought entered his head which he seemed to have had before. It was something he had heard, felt, and thought in the past. . . . He was unable to describe it."

SZ: "Suddenly, the patient experiences a sensation of recollection, which feels like a scene she had experienced somewhere in the past. She feels as if she has seen something familiar. As she tries to recall what it is, she feels a sense of pleasure."

The authors comment: *"In this description, the familiarity is disso-ciated from memory and 'the feeling of knowing' appears in mind."*[21]

Jamais Vu and Other "Feelings of Strangeness"

CS: "I had a dream—I wasn't here. . . . I sort of lost touch with re-ality. . . ." Stimulation was repeated at the same site. "A small feel-ing like a warning." The stimulation was again repeated. "I was *losing touch with reality* again."

SZ: "He had a sensation of *'strangeness of words'* as if he had never seen or heard them before."

SZ: "His aura begins with a sense that *objects look bizarre*, and that speech, although understood, *sounds strange* in an indefinable way."

CS: *"Things are deformed. . . . I am another person* and *I seem to be somewhere else."* The patient also described anguish with a feel-ing of imminent death.

CS: "From the age of thirty-five, the patient has suddenly and transiently felt as if she were *falling into another, and fearful, world.*"

CS: "He felt himself *alone in another world,* and he felt fearful."

SZ: "When he has an attack in his own room, he feels as if *his room has been changed and has become strange.*"[22]

Strangely Familiar—a Duet of Opposites

Descriptions that include simultaneous feelings of familiarity and strangeness:

SZ: "A brief 'dream' without loss of consciousness, where sud-denly he had a very strong memory of a scene that he has already lived through, that nonetheless *feels bizarre.* Later, the scene was preceded by 'the impression of having already done what I am in the process of doing; it seems to me that I have already lived through the entire situation; with a *feeling of strangeness* and often of fear.' "

SZ: "Begins with a feeling of fear, then an indefinable *internal feeling of strangeness, sometimes associated with the emergence of old or recent memories* (presented more as thoughts than as sensory images)."

SZ: "Begins with a *very agreeable aesthetic illusion. . . .* that would appear to him as if it were *magnificent,* giving him *great pleasure.* At about the same time, *intense thoughts would come to him, which he would accept uncritically;* it could be a voice, like in a dream—*he thinks that someone wished him harm, that people are saying bad things about him, but at the same time he takes pleasure from this.*"

SZ: "Begins with a feeling of discomfort and epigastric constriction; a *feeling of strangeness and unreality of the environment,* with a vague feeling of *déjà vécu;* then loss of contact.

"I am in a small village where everyone knows each other. . . . I had the impression of having seen those people, and I felt something in the stomach, like a ball . . . that which I saw, could have been anything. It's more like an idea than an image that was presented rapidly in a very fleeting manner; *some strange thing, without relation to reality,* moving, but not necessarily, with lifelike colors."

SZ: "Begins with an indefinable feeling of fear, sometimes associated with an internal whispering voice and then an *intensely painful emotional state with a familiar resonance, 'like the memory of an emotion.'* "

MY AIM IN presenting these detailed descriptions is not to categorically identify the limbic system as the sole site of origin of the feelings such as familiarity, realness, "knowing," clarity of thought, and so on, but to show how these feelings that qualify how we experience our thoughts can be elicited both chemically and electrically without any antecedent triggering thought or memory. Familiar and real aren't conscious conclusions. Neither are strange and bizarre. They are easily elicited without any associated reasoning or conscious thought. But what exactly are these "mental states"? A word of clarification is in order.

4

The Classification
of Mental States

The fish trap exists because of the fish. Once you've gotten the fish, you can forget the trap. The rabbit snare exists because of the rabbit. Once you've gotten the rabbit, you can forget the snare. Words exist because of meaning. Once you've gotten the meaning, you can forget the words. Where can I find a man who has forgotten the words so that I can talk with him?

—Chuang-Tzu (c. 200 B.C.E.)

WITTGENSTEIN'S OBSERVATION THAT DIFFICULTIES IN PHI-losophy eventually boil down to problems with language surely applies to the study of the mind. The present-day classifications of mental states are a huge obstacle to any deeper understanding of how the mind works. While fear is obviously an elemental emotion, feelings from tip of the tongue to utterly strange, from totally real to otherworldly, are neither pure emotions nor thoughts. They are feeling tones that color our mental experiences. In writing this book, I have struggled with what to call these feelings, and have come up short. Ideally, the label should be an accurate reflection of the underlying physiology.

The behavioral neurologist Antonio Damasio sums up our present state of ignorance. "Deciding what constitutes an emotion is not an easy task, and once you survey the whole range of possible phenomena, one does wonder if any sensible definition of emotion can be formulated, and if a single term remains useful to describe all these states. Others have struggled with the same problem and concluded that it is hopeless."[1]

Psychologists commonly divide certain feeling states into *primary emotions*, such as happiness, sadness, fear, anger, surprise, and disgust,[2] and *secondary* or *social emotions*, such as embarrassment, jealousy, guilt, and pride.[3] Methods of classifications and the number of primary emotions vary depending upon what is being measured—from universal facial expressions or basic motor activity to the language used when we talk about emotions.[4]

No one would question that embarrassment is a sickening feeling and in common parlance is a full-fledged emotion manifested by obvious behavioral correlates such as a flushed face and a glance toward the nearest exit. But what about those so-called emotions that are so devoid of emotional tone that they feel more like thoughts? Is gratitude an emotion, a thought, or a shifting combination dependent upon yet other moods? Each morning, I think about how fortunate I am. I tell myself to be grateful, and I am. For me, gratitude feels like a comparison—an intellectual exercise rather than an emotion (though it might result in a sense of contentment, which is more of a mood than a raw emotion). I am thankful for X, which implies being better off than Y.

I cannot imagine feeling proud without being proud *of something.* We take pride in, are grateful for, embarrassed at, or are pleased with—to my knowledge, these more complex emotional states rarely, if ever, occur with complex partial seizures or brain stimulation studies, unless coincidentally provoked by an accompanying thought or memory. There is no body of neurological literature demonstrating the isolated absence of pride or gratitude as the result of localized brain injuries. Such emotions do not appear to be primary any more than magenta is a primary color. They are the end product of other more elemental mental states.

What about the other so-called primary emotions? Surprise is a spontaneous response to the unexpected. But being surprised by a good punch line to a joke or a great twist in a thriller requires some cognitive element. (I am using the term *cognitive* to refer to any form of thought, conscious or unconscious, as opposed to a feeling, mood, or emotion.) You expect one thing and experience something different. Also, a sense of surprise is not easily elicited by brain stimulation. And happiness? Is this an emotion or a mood? One person's happiness is another's relief that things aren't worse.

If emotions as basic as surprise are difficult to physiologically categorize, what would a reasonable approach be to the even more elusive *feeling of knowing*? Perhaps an analogous situation would be the stream-of-consciousness voice in my head. Though not audible, I do "hear" my internal voice in the same way as I "see" an object in my mind's eye. Both are sensory representations of internal states of my mind.[5] So is the *feeling of knowing*. We perceive our external world through primary senses such as sight, sound, and smell; we perceive our internal world through feelings such as familiar or strange, real or unreal, correct or incorrect, and so on.

Earlier, I mentioned that déjà vu could be described as a mental sensation. The term *mental sensation* is stilted and unwieldy; *feeling* still feels (!) preferable. Yet there are several compelling reasons for at least considering these feelings as sensations—as in a *sense of conviction.*[6] Sensation strikes closer to the neurophysiological truth of a relatively discrete output from localized neural structures in the same way that vision is the sensory output of the eye and its related cortical areas of the brain. Sensation tends to minimize the emphasis on psychological factors; talk of feelings encourages this relationship.

But the most practical reason to consider these mental states as sensations is that they are subject to certain physiological principles common to other sensory systems. If you cut the median nerve—the main sensory nerve to your thumb—you cannot volitionally stop your thumb from feeling numb. When a sensory system has been affected, altered sensations are unavoidable. Similar phenomena occur within the brain. Consider the example of phantom limb pain. A man's arm is accidentally amputated. The region of the brain that previously received sensory inputs from the now missing arm undergoes changes that cause it to misfire. The tragic result is the ghostly and often painful re-creation of the missing arm—the so-called phantom limb. As with the median nerve injury where you cannot will away the numbness, the amputee can clearly see that the arm is missing, yet cannot stop the disturbing phantom limb sensations.

The same line of reasoning can allow us to rethink such bizarre beliefs as feeling that you are dead or that your prized antique desk is a cheap reproduction. We wouldn't expect that such beliefs—if the product of altered mental sensory systems—could be voluntarily overcome through reason or contrary evidence. The same logic also applies to the *feeling of knowing.*

During the height of his mental illness, Nobel Prize–winning mathematician John Nash believed that aliens from outer space were trying to communicate with him. He could not accept a full professorship at MIT because "I am scheduled to become the emperor of Antarctica."[7] When a colleague asked him how such a brilliant and logical man could believe such nonsense, Nash replied that both ideas had come to him in the same way. Both thoughts *felt right.*[8]

And the converse: If you've known someone with severe obsessive-compulsive disorder (OCD), you have seen how they cannot rely upon what they should know to be true. They will repeatedly check the oven to be sure the gas is off, triple-check the locks that they can easily see are already locked, or count and recount their change. It is as though objective evidence cannot trigger a proper *feeling of knowing,* leaving OCD victims in a state of heightened doubt and anxiety.

Psychologists have recently begun to consider the role of *pathological certainty* and *pathological uncertainty* as they relate to schizophrenia and OCD.[9] This is an intriguing possibility given that both mental disorders have a significant genetic contribution. Could genetic differences play a role in how easily one becomes "convinced" or remains "unconvinced"? Might inherent variations in the expression of the *feeling of knowing* contribute to the characterological excesses of the know-it-all, the perennial skeptic (the doubting Thomas), or the patient with psychosomatic complaints who is sure that something is wrong despite negative tests? But I get ahead of myself.

A CLASSIFICATION OF mental states might also be based upon the degree of neurological autonomy. The most elemental would be those feelings and emotions that are universal, deeply imbedded in our neural circuitry, and capable of activation independently of

any thought. Further confirmation would be the demonstration
of a relatively specific site of origination, as in the well-established
relationship between fear and the amygdala, or déjà vu and the
temporal lobe.

More complex states are those that tend not to be well-
localized, or spontaneously arising, and require some contribution
from thought, memory, or conscious intervention. Déjà vu is universal and spontaneous; guilt and regret are not (ask Bill Clinton
or Dick Cheney). Brain stimulation experiments or discrete brain
lesions do not generate a disparaging internal commentary, an isolated sense of pride, indignation, guilt, or humility. No one is on
record as having experienced a sense of irony as a primary epileptic aura. Some brain activities such as hope seem to defy categorization altogether.

To summarize: By using these criteria of universality—relatively
discrete anatomic localization and easy reproducibility without
conscious cognitive input—the *feeling of knowing* and its kindred
feelings should be considered as primary as the states of fear and
anger. The recently defined relationship between fear or anxiety
and conscious thought has spawned the concept of *emotional intelligence;* it is time for a similar examination of the role of the
feeling of knowing in shaping our thoughts.

5

Neural Networks

Essentially everything that the brain does is accomplished by the process of synaptic transmission.

—Joseph LeDoux, *Synaptic Self*

IF THE *FEELING OF KNOWING* IS A PRIMARY MENTAL STATE not dependent upon any underlying state of knowledge, then our next step is to see how the interaction of conscious thought and the involuntary *feeling of knowing* determines how we feel we know what we know. We needn't fret over the enormously complex details of the underlying neurobiology; what is important is a good grasp of the key elements governing brain hierarchical structure. By understanding how increasingly complex layers of neural networks emerge seamlessly into the conscious mind, we will have the foundation for seeing where contradictory aspects of thought collide and why certainty is contrary to basic biological principles. In this chapter, we will take a look at neural networks.

In the human brain, a typical neuron receives incoming information from approximately ten thousand other neurons. Each bit of information either stimulates (positive input) or inhibits (negative input) cell firing. The neuron acts like a small calculator. If

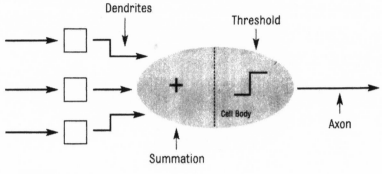

Dendrites

Threshold

Cell Body

Axon

Summation

A neuron

the sum of the inputs reaches a critical threshold level, an electrical charge travels down the nerve fiber (axon) to the region where neurotransmitters are stored. The transmitters are released into the synaptic cleft—a tiny gap between adjacent neurons. If a neurotransmitter finds a receptive site (receptor) on the adjacent neuron, the process will be repeated on this adjacent neuron.

Every step in the process of neuronal activity—from the most distant dendrite to the farthest axon terminal—is fine-tuned by a slew of control mechanisms. There are estimated to be at least thirty separate neurotransmitters with enzymatic steps in the creation and destruction of each transmitter affected by everything from genetics to disease. Feedback loops alter the availability and receptivity of postsynaptic receptor sites and even how cells signal and adhere to one another. (Understanding these regulatory mechanisms is a major challenge of modern neurobiology.)

Despite a veritable symphony of interacting mechanisms, the neuron ultimately only has two options—it either fires or it doesn't. At this most basic level, the brain might appear like a massive compilation of on-and-off switches. But the connections

A synapse

At a presynaptic terminal (top) small vesicles twenty to thirty nanometers in diameter filled with neurotransmitter molecules are waiting. Arrival of an action potential (or spike) induces a fusion of the membrane with some of the vesicles so that a neurotransmitter can diffuse into the synaptic cleft and reach receptors (not shown) at the other side, which then open ion channels they are attached to. A synapse becomes more or less efficient, like when vesicles get bigger or smaller, or more or fewer release sites become available, while postsynaptically the ion channels may increase or decrease in number and stay open during a longer or shorter period of time. So most, though not all, of the active processes are happening in the pre- and postsynaptic membrane. The result is called learning.

Image courtesy of the Synaptic Corporation, Aurora, Colorado, United States; www.synapticusa.com.

between neurons are not fixed entities. Rather they are in constant flux—being strengthened or diminished by ongoing stimuli. Connections are enhanced with use, weakened with neglect, and are themselves affected by other connections to the same neurons. Once we leave the individual synapse between two neurons, the complexity skyrockets—from individual neurons to a hundred billion brain cells each with thousands of connections. Although unraveling how individual neurons collectively create thought remains the Holy Grail of neuroscience, the artificial intelligence (AI) community has given us some intriguing clues as to how this might occur.

Using the biological neuron and its connections as the model, AI scientists have been able to build artificial neural networks (ANN) that can play chess and poker, read faces, recognize speech, and recommend books on Amazon.com. While standard computer programs work line by line, yes or no, all eventualities programmed in advance, the ANN takes an entirely different approach. The ANN is based upon mathematical programs that are initially devoid of any specific values. The programmers only provide the equations; incoming information determines how connections are formed and how strong each connection will be in relationship to all the other connections (or weightings). There is no predictable solution to a problem—rather as one connection changes, so do all the others. These shifting interrelationships are the basis for "learning."

The AI community has labeled this virtual space where the weightings take place as the *hidden layer.*

With an ANN, the *hidden layer* is conceptually located within the complex interrelationships between all acquired (incoming) information and the mathematical code used to process this information. In the human brain, the *hidden layer* doesn't exist as a discrete interface or specific anatomic structure; rather it resides within the connections between all neurons involved in any neural network. A network can be relatively localized (as in a specialized visual module confined to a small area of occipital cortex), or can be widely distributed throughout the brain. Proust's taste of a madeleine triggered a memory that involved visual, auditory, olfactory, and gustatory cortex—the multisensory cortical represen-

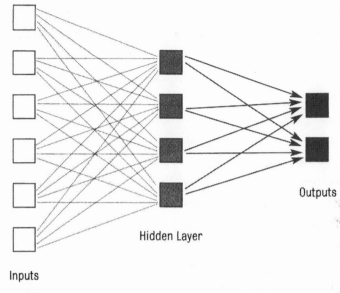

Outputs

Hidden Layer

Inputs

A neural network

tations of a complex memory. With a sufficiently sensitive fMRI scan, we would see all these areas lighting up when Proust contemplated the madeleine.

The *hidden layer,* a term normally considered AI jargon, offers a powerful metaphor for the brain's processing of information. It is in the hidden layer that all elements of biology (from genetic predispositions to neurotransmitter variations and fluctuations) and all past experience, whether remembered or long forgotten, affect the processing of incoming information. It is the interface between incoming sensory data and a final perception, the anatomic crossroad where nature and nurture intersect. It is why your red is not my red, your idea of beauty isn't mine, why eyewitnesses offer differing accounts of an accident, or why we don't all put our money on the same roulette number.

I have borrowed the term *hidden layer* from the AI community to highlight a crucial element of neurophysiology. All thought that manipulates ideas and information by shifting associations (relative valuations) among myriad neural networks must also arise from these hidden layers.

Because the hidden layer is such an important concept, let's follow the inner workings of an ANN familiar to most of us: The book recommendations on Amazon.com. Anyone who has shopped on Amazon more than once has had the disquieting experience of having the Web site suggest additional books that you might enjoy. The software advising you is an ANN program that compiles a database of all the book sites on Amazon that you visit. The first time that you log on to Amazon, there are no recommendations. The ANN has no idea of your preferences. Though the mathematical equations are in place, they are useless without your input.[1] Then you begin surfing the site. Each click onto a book inputs information into the ANN database. Gradually a pattern develops; books become ranked in relation to each other (weighting), depending upon whether you clicked onto the book only, pursued reading a sample chapter, or purchased the book. Obviously, for Amazon, a purchase will be more heavily weighted than a rejection after perusing a sample chapter.

In effect, the ANN learns your preferences and which books, if recommended, you are most likely to buy. The ANN has formed the equivalent of neural links between your initial purchases and similar books at Amazon. If, when you first started using Amazon, you only searched for and bought crime novels, further suggestions would be primarily in this genre, with some overlap to the most closely related areas, like true crime or biographies of Sherlock Holmes. The more crime books you buy, the more the underlying neural network would be weighted toward recommending similar books.

Then your wife fires a volley of disparaging comments about your reading tastes. After some reluctant self-examination, you glumly concur. You agree to a moratorium on loading up on pulp fiction. Instead, you will only read existential philosophy and plays from the theatre of the absurd. You click on Pinter and Beckett, and order a copy of *Waiting for Godot*. The next time that you boot up Amazon.com, you will still get crime novel recommendations, but at the bottom of the list is a recommendation for Camus's *The Plague*. Sounds a bit like a thriller, so you order the book.

The next time that you sign on to Amazon, there are recommendations for books by Sartre and Ionesco. Elmore Leonard's latest is further down the list. If you stop reading crime novels long enough, the weightings of crime novels within the database will gradually revert toward zero. In essence, the program is learning your tastes by keeping a detailed track of what you read/don't read and purchase/don't purchase. It is building a relational database—one that is continuously adjusted according to new experience (if you can say that a database is having an "experience"). If you like the hard-boiled dialogue of Raymond Chandler, it would seem logical that you would be more likely to appreciate Jim Thomson's *The Grifters* than if you preferred Henry James's prose. If so, some static algorithmic program might be able to make preprogrammed recommendations. But line-by-line programming cannot mimic the inconsistencies and unpredictable nature of taste. It will continue to give the same recommendations until it is rewritten.

By contrast, the ANN is continuously learning from its mistakes. It can monitor its recommendations by accessing your purchases. If it is right—if you buy both James and Leonard despite their apparent differences—the ANN will get immediate feedback about your idiosyncratic aesthetics. Subjectivity, whimsy, and all sorts of unpredictable correlations will be included in these weightings. Even the purchases of others affect the weightings. If one thousand Elmore Leonard readers suddenly buy a Danielle Steel novel, you might become bombarded with recommendations for her latest romance.

If we were to envision each book on Amazon as a neuron connected to all the other available books (neurons), we would have the beginning model of a neural network. How a book relates to another book is being constantly recalculated (reweighted) based upon the shifting relationships among all of the books.

An important conceptual point: The reader can keep track of which books he has clicked onto and tabulate his inputs. He can record the recommendations made by Amazon (outputs). But the world's smartest AI consultant cannot tell him in advance why the ANN acted as it did. There is no underlying program or algorithm that contains a reason. The process depends upon the entire set of interrelationships, none of which are fixed. One cannot extract a piece of the network for independent observation any more than you can pull out a single strand of a Persian rug and infer what the rug's pattern might be.

Here is the sequence of events:

INPUT: clicking onto a book at Amazon.com.

THE HIDDEN LAYER: weighting of relationships among all books, clicked on or purchased, which occurs within the interdependent formulations that comprise the "guts" of the neural network.

OUTPUT: recommendation for additional purchases.

The simplest neural network involves a single input and a single output. More complex networks result from multiple inputs and multiple outputs.

NOW LET'S UP the ante and watch a human neural network in action. A bright light is briefly flashed into your eyes. The retina turns the flash of light into electrical data that travel along the optic nerves and into the brain (input). But instead of a direct route to consciousness with a precise and unaltered duplication of the flash, the data first goes to a subconscious holding station where it is scrutinized, evaluated, and discussed by a screening committee representing all of your biological tendencies and past experiences. This committee meets behind closed doors, operating outside of consciousness in the hidden layer.

Consider each committee member as being one set of neural connections. One might represent a childhood memory of having seen a similar flash of light when a toaster shorted out and started an electrical fire; the second is a general alarm system that has recently become highly sensitive and vigilant to the possibility of terrorism; the third is a composite memory of rock concerts; the fourth is a genetically based predisposition for a heightened startle reflex for bright lights. Each member has his own opinion and each gets one vote. After hearing all the arguments, each committee member casts his vote and they are tallied (weighted). At the most elemental level, a decision is made—either to entirely suppress the flash or send it on to consciousness (output). The degree of awareness generated is yet another function of this decision—ranging from a barely noticed flash at the periphery of vision to a bright flash, front and center.

The childhood memory votes yes: Send the flash into awareness. The terrorist alarm network, fearing that the flash could indicate an explosion, votes yes. The rock concert memory is blasé, has seen the same flashes a zillion times at rock concerts, and feels the flash should be ignored. It votes no. The genetic predisposition reflexively votes yes.

The third member is outvoted, and the flash is sent on high priority into consciousness. You look around, heart pounding, on high alert for everything from a gunshot to a terrorist bomb exploding. But you are at a wedding, and everyone is taking pictures of the bride. You sigh and tell yourself not to be so anxious.

The next time a similar flash is received, the committee reminds each of the members that last time was a false alarm. Some of the committee members who previously voted yes now feel sheepish and don't vote. The committee votes to nearly completely suppress the image. The genetic predisposition is ignored. So you barely notice the flashbulbs going off while you watch your child play Elmer Fudd in his grammar school play.

Eventually, if the committee is presented with the flash enough times, and there is no explosion or fire, even the most nervous committee members reluctantly give up their alarmist posture. At this point, you could say that the neural network was heavily weighted toward suppression of the incoming flash. Unless there was a subsequent alternative outcome such as a fire or an explosion, the vote would evolve into a rubber stamp veto. Professional photographers pay no attention to other flashing cameras (unless they think that they are getting scooped).

In this schema, each committee member represents a neural network with his own particular bent or bias. With the possible exception of some hardwired genetic tendencies, each member is also capable of listening to and being influenced by other networks. If he likes another network's view of the world, he might reach out and join ranks by increasing (enhancing) the connections with this other network. Conversely, if he does not, he may reduce his connections with the offensive member (network). It is not possible to know how each member is going to respond without knowing how each of the other members will respond. Everyone is constantly watching everyone else, with each decision being influenced by what the others are doing.

To get an idea of the magnitude of this process, imagine billions of committee members, each with at least ten thousand hands reaching out to shake hands, prod, poke, seduce, or fend off the other members. Miraculously, this orgy of utter chaos is transformed into a relatively seamless and focused stream of consciousness. Even given the amount of potential information incoming at any instant, we can focus on a single aspect of consciousness and either not notice or ignore the enormous subconscious din.

The schema of the hidden layer provides a conceptual model of a massive web of neuronal connections microscopically interwoven throughout the brain. Such neural networks are the brain's real power brokers, the influence peddlers and decision makers hard at work behind the closed doors of darkened white matter. How consciousness occurs remains an utter mystery, but conceptually, it must arise out of these hidden layers.

The concept of neural networks also helps explain why established habits, beliefs, and judgments are so difficult to change. Imagine the gradual formation of a riverbed. The initial flow of water might be completely random—there are no preferred routes in the beginning. But once a creek has been formed, water is more likely to follow this newly created path of least resistance. As the water continues, the creek deepens and a river develops.

On your first visit to Amazon, you have no particular preference in mind. You randomly pick out a bestseller—an Elmore Leonard novel. The next time you click on Amazon, you will be bombarded with recommendations for other crime novels. Perhaps you hadn't planned on buying another, but you are seduced by the blurbs and reviews. Eventually, your recommendations are a reflection of a chance initial purchase—like the beginning of a stream that mindlessly deepens itself.

The brain is only human; it, too, relies on established ways. As interneuronal connections increase, they become more difficult to overcome. A hitch in your golf swing, biting your nails, persisting with a faulty idea, not dumping your dot.com stocks in late 1999—habits, whether mental or physical, are exasperating examples of the power of these microscopic linkages. At the most personal level, most of us glumly acknowledge that we could abandon many of our failed self-improvement efforts if we could somehow painlessly alter these neural networks. Yet B. F. Skinner was roundly booed for pushing behavioral modification. (If he'd had his way, we'd have been raised like veal.) But he was not alone. The idea of somehow undoing the circuitry is not just the stuff of science fiction; it is a recurring theme in medicine.

In 1935, Egas Moniz, a Portuguese neurologist and Nobel laureate, observed: "It is necessary to alter these synapse adjustments and change the paths chosen by the impulses in their constant passage so as to modify corresponding ideas and force thoughts into different channels. . . . By upsetting the existing adjustments and setting in motion other [connections], I [expect] to be able to transform the psychic reactions and to thereby relieve the patients."[2]

In 1936, Dr. Moniz introduced a surgical procedure—prefrontal leucotomy—primarily for the treatment of schizophrenia. The operation—later referred to as a frontal lobotomy—was designed to destroy connections between the prefrontal region and other parts of the brain. In 1949, the Nobel committee said of Moniz's work, "Frontal leucotomy, despite certain limitations of the operative method, must be considered one of the most important discoveries ever made in psychiatric therapy."[3]

What is remarkable about Moniz was his prescience in predicting the power of neural networks combined with a profound naïveté in believing that they could be surgically altered. If you want to see how these patients turned out, watch George Romero's *Night of the Living Dead* or Miloš Forman's *One Flew Over the Cuckoo's Nest*. (In fairness to Moniz, at that time medicine had little else to offer the severely psychiatrically disturbed. The era of modern psychopharmacology began with the introduction of the first phenothiazine, Thorazine, in 1954.)

But old ideas die hard—perhaps because the way we conceptualize the practice of medicine is itself a pattern difficult to change. Hence the continuing surgical mentality of "if it seems diseased, cut it out." The head of stereotactic and functional neurosurgery at the Cleveland Clinic has recently suggested that a combination of microneurosurgical techniques, implanted computer processors, and evolving molecular biological strategies might be able to "replace entire neural networks that become affected by psychiatric and other neurological diseases."[4]

Is this science fiction, wishful thinking, utter madness, or a heartfelt and genuine attempt by neuroscientists struggling to mold these essentially infinite connections into a workable medical model? Networks aren't localized like a spot of rust on a fender. They aren't separable into their component parts any more than a cake can be reverse engineered into eggs, sugar, flour, water, and chocolate. These networks *are* the brain.

6

Modularity and Emergence

Organizing Complexity

Seeing each individual neuron as a simple on-and-off "device" is convenient yet profoundly deceiving. The final yes-no decision to fire or not to fire is influenced by complex control mechanisms ranging from the interactions of genes to moment-to-moment shifts in hormone levels. Understanding how the mind works would require nothing less than a full understanding of these relationships at every instant *and* the ability to accurately predict the final output of such competing forces. Once we leave the individual neuron, the scale of interaction becomes exponentially more complex. Fortunately, for the purposes of our discussion, we don't need to get bogged down in endless speculations and moment-by-moment updates of what is presently known about such mechanisms. To understand the origins of a thought, we can get by with the grossest of simplifications: Individual "mindless" neurons join together to mysteriously create the mind. Which brings us to the interrelated concepts of modularity, hierarchical structure, and emergence.

Modules

Perhaps the most intensively studied region of the brain—the visual system—provides us with an excellent generalization about how the brain converts lower-level functions to higher-level behaviors. The visual cortex is organized into clusters of cells that selectively respond to the various components of vision, from the recognition of discrete angles, lines, and edges, foreground and background, to the detection of motion and color. Such neurons only fire when presented with certain categories of stimuli, but not with others. For example, a cell might respond maximally to one angle of light, but less so for other angles of light, and not at all for yet others. A cell might respond to a particular shape but not another. A cluster of these highly individualized neurons specific to a single visual function is referred to as a *module*.[1]

The Hierarchical Arrangement of Sensory Data

Your retina detects an orange-and-black fluttering. The information is sent to the primary visual cortex. Each category of module gathers its own particular data (such as the detection of vertical or horizontal motion, color, shape, and size). No single module can create a visual image. Rather the output of each flows into higher order networks within the visual association areas where it merges with a host of inputs from nonvisual circuitry—the remembrance of seeing a similar pattern hovering over a mountain lake, a trip to a natural science museum with your grandfather, the cover of a book on chaos theory, a scary scene from *Silence of the Lambs*. The sensory detection committee within the visual association hidden layer weights the inputs and casts its ballot; the output becomes the perception and recognition of a monarch butterfly hovering on your front porch.

Modules process different aspects of vision, yet work as a team. We can't see pure motion despite having a module for motion detection. We need to see an object or shape moving. Similarly, we can't see pure color in the absence of some form. Awareness of individual modules occurs primarily when they fail to operate properly, leaving a hole in our fabric of perception. For example, a small stroke limited to the area of the occipital cortex that controls movement detection can cause a sudden inability to see images in motion. One such patient reported seeing a stalled car on the road some distance away; then, while continuing to observe the car, it suddenly was looming directly in front of her. Unable to detect motion, she saw only a succession of discontinuous still shots of the car. When pouring a cup of tea, she saw a frozen arc of tea rather than flowing water. Only when she saw a puddle appear on the floor did she realize that the cup had overflowed. She was neither able to see the car approaching nor the cup filling up.[2] By working backwards from such case histories, neurologists have been able to identify at least thirty discrete modules that generate the visual image (although it is likely that we will uncover others).

Modules are the building blocks of perception, but are not normally individually detectable. An aside: Be thankful that we do not normally experience the separate effect of each module that contributes to a visual image. Being constantly aware of the scaffolding of perception would be frustrating and confusing, the incoming sensory information would be as unnecessary as labeling every item we see or touch. Imagine a world in which you had to eat the recipes along with the meal.

The stripped-down model of brain hierarchy is that the individual neurons, which contain no imagery and operate outside of awareness, flow into progressively higher-order networks until a picture emerges. In AI models—which are extraordinarily simple in relationship to the most primitive animal brain—the conversion of lower-level information into the final image is accomplished via a series of mathematical calculations within the *hidden layer* of the neural networks. The precise mechanisms remain a profound mystery and the key to understanding how consciousness arises out of "mindless" neurons. To give this extraordinary process a commonsense explanation, scientists have provided us with the self-defining, yet intuitively appealing, theory of *emergence*.

Emergence

A classical example of emergence is how termites with their tiny brains are able to construct huge mounds up to twenty-five feet in height. No termite has a clue how or why to build a mound; its brain isn't large enough to carry the information. There are no termite engineers, architects, or critics; all termites are low-level laborers operating without blueprints, or even a mind's eye notion of a termite mound. Yet the mound is built. Somehow the interaction of lower-level capabilities produces a higher-level activity.[3]

The same process applies to the human brain. Each neuron is like a termite. It cannot contain a complete memory or hold an intelligent discussion. There are no superneurons, nor is there a master plan contained within each neuron. Each neuron's DNA provides general instructions for how a cell operates and relates with other cells; it does not provide instructions for logic, reason, or poetry. And yet, out of this mass of cells comes Shakespeare and Newton. Consciousness, intentionality, purpose, and meaning all emerge from the interconnections between billions of neurons that do not contain these elements.[4] Termites are to termite mounds as single neurons are to the mind. Primary modules provide the *bricks and mortar*, the secondary association areas build the *house*, and yet more complex interactions are necessary to call this building *home*.

Modularity, when combined with a schematic hierarchical arrangement of increasingly complex layers of neural networks and the concept of emergence, serves as an excellent working model for how the brain builds up complex perceptions, thoughts, and behavior. Harvard psychology professor Steven Pinker has even gone so far as to suggest that we use the term *modules* interchangeably with "mental organs" to emphasize that the brain is composed of many functionally specialized mechanisms that collectively create the "mind." Of course this isn't literally true—the brain is a single organ—but it does help us conceptualize how aspects of complex behaviors can be broken down into more manageable bits. The bad news and a huge caveat when applying the idea of modularity to behavior is that excessive reductionism or a fuzzy definition of a behavior can lead to grand nonsense. Anyone familiar with the biographies of Rockefeller, Carnegie, and Gates realizes that acts of charity cannot be easily attributed to precise motivations and urges or a gene for altruism. One man's compassion is another's tax deduction.

A quick word about module localization. When neurologists talk of visual modules, they are referring to columns of adjacent cells within the visual cortex that perform the various tasks necessary to create a visual image. These modules are anatomically discrete, confined to a small region of the brain, and can be identified with standard neurophysiological studies such as microelectrode intracellular recordings. Modules for behavior aren't well-localized; they represent widely distributed aspects of a common function. Steven Pinker's wonderfully disgusting description bears repeating.

> The word "module" brings to mind detachable, snap-in components, and that is misleading. Mental modules are not likely to be visible to the naked eye. . . . A mental module probably looks more like roadkill, sprawling messily over the bulges and crevasses of the brain. Or it may be broken into regions that are interconnected by fibers that make the regions act as a unit.[5]

Consider the myriad components involved in the acquisition of language, ranging from the visual recognition of symbols and auditory processing of spoken sounds (phonemes) to the sorting out of nuance and implied irony. A racial epithet can be an accusation or a term of endearment, depending upon circumstance, facial expression, body language, and intonation. (Comedian George Carlin has made a career out of forcing us to hear politically loaded words from unanticipated angles.) In the interpretation of a single word, large areas of widely separated but interconnected cortex function as a behavioral unit—hence the applicability of the term *module*.

Any classification of widespread nondiscrete modularity at this level requires a leap of faith—that these anatomically separate areas of brain contributing to a behavior are actually genetically linked in the same way that a Tinkertoy's components are part of a grand toy design. In time, it is likely that some behavioral traits will meet such criteria; others will be delegated to the trash heap of outdated psychology. Nevertheless, some version of modularity is essential to any unraveling of the biology of behavior. Whether talking about genes for risk-taking, perfect pitch, or mathematical ability, or the adaptive value of compassion, deceit, or Machiavellian cunning, evolutionary biologists take as their starting point the assumption that certain biological attributes are integral to the expression of behavioral traits. If evolution is responsible, genetic transmission is presumed.

The general concept of modularity is a powerful tool for generalizing how the brain functions, including the formation of our thoughts. The *feeling of knowing* is universal, most likely originates within a localized region of the brain, can be spontaneously activated via direct stimulation or chemical manipulation, yet cannot be triggered by conscious effort. These arguments for its inclusion as a primary brain module are more compelling than those postulated for deceit, compassion, forgiveness, altruism, or Machiavellian cunning. One can stimulate the brain and produce a *feeling of knowing;* one cannot stimulate the brain and create a politician.

What a predicament. The idea of a thought being created by more specialized modules, some operating outside of our control and awareness, seems both intuitively obvious and antithetical to how we experience our thoughts. I am not talking about the difference between conscious and unconscious cognition, but am referring to how we build a thought from "scratch." At stake is the concept of a rational mind. To begin this exploration, we should take a page from the neurologist's approach to brain function and look for conditions in which an inappropriate activation of a module affects thought in unanticipated and unintentional ways. One of the most fascinating and insightful is the phenomenon of synesthesia.

Synesthesia

First described in 1880 by Francis Galton, a cousin of Charles Darwin, synesthesia is commonly thought to represent an involuntary comingling of two normally unrelated sensory modalities, such as sight and sound. Those affected experience two separate sensations as a single unit; they cannot willfully suppress the second sensory input. A synesthete might hear colors, taste shapes, and describe the color, shape, and flavor of someone's voice or music. According to those neurologists who have spent considerable time interviewing synesthetes, these perceptions are experienced as "real" as opposed to mere illusion, hallucination, or being seen in the "mind's eye."[6]

Here are two typical descriptions. A forty-seven-year-old psychologist: "New Orleans-type jazz hits me all over like heavy, sharp raindrops. The sound of guitars always feels like someone is blowing on my ankles." Patricia Duffy, a journalist and cofounder of the American Synesthesia Association: "I can't remember his name, but I know it's purple."[7]

Vladimir Nabokov on the alphabet:

The long a of the English alphabet . . . has for me the tint of weathered wood, but a French a evokes polished ebony. This black group also includes hard g (vulcanized rubber) and r (a sooty rag being ripped). Oatmeal n, noodle-limp l, and the ivory-backed hand-mirror of o take care of the white. . . . Passing on to the blue group, there is steely x, thundercloud z and huckleberry h. Since a subtle interaction exists between sound and shape, I see q as browner than k, while s is not the light blue of c, but a curious mixture of azure and mother-of-pearl.[8]

Arthur Rimbaud, Wassily Kandinsky, Vladimir Nabokov, David Hockney,[9] and Alexander Scriabin are just a few of the great artists who have had this eerie ability, if *ability* is the proper word. To see how these involuntary sensations shape both behavior and thought, here's David Hockney's description of creating a set design for the Metropolitan Opera: "I listened to the Ravel music and there's a tree in one part of it, and there's music that accompanies the tree. When I listened to that music, the tree just painted itself." For Hockney, the musical sound of a segment of Ravel triggered his brain to "see a tree." (Hockney also has spoken of hearing the colors that he has painted.)[10]

Alexander Scriabin, the Russian composer and pianist, was one of the first synesthetes to thoroughly catalogue his color-musical note associations. C-sharp was violet and E was pearly white and the shimmer of moonlight.

Neurologist V. S. Ramachandran offers some compelling speculations as to how synesthesia might occur. "Perhaps a mutation causes connections to emerge between brain areas that are usually segregated. Or maybe the mutation leads to defective pruning of preexisting connections between areas that are normally connected only sparsely." Though Ramachandran initially thought in terms of physical cross wiring, he now believes that the same effect could also occur with neurochemical imbalances between regions. "For instance, neighboring brain regions often inhibit one another's activity, which serves to minimize cross talk. A chemical imbalance that reduced such inhibition—for example, by blocking the action of an inhibitory neurotransmitter or failing to produce an inhibitor—would also cause activity in one area to elicit activity in a neighbor. Such cross activation could, in theory, also occur between widely separated areas, which would account for some of the less common forms of synesthesia."[11]

Synesthesia commonly runs in families; most neurologists accept a genetic component. A journal entry from Carol S, a New York artist: "I was sitting with my family around the dinner table, and I said, 'The number five is yellow.' There was a pause, and my father said, 'No, it's yellow-ochre.' . . . At that time in my life I was having trouble deciding whether the number two was green and the number six blue, or just the other way around. And I said to my father, 'Is the number two green?' and he said, 'Yes, definitely. It's green.' "[12]

An interesting sidebar: Members of a family do not necessarily experience the same colors, or even the same types of synesthesia. The same presumed gene (or genes) can produce similar or dissimilar experiences—yet another argument for distinguishing between genes associated with a particular behavior and the actual manifestation of the behavior.

Private Islands

Synesthete Patricia Duffy elegantly summarizes how these differences in perception are at the heart of different worldviews.

In life, so much depends on the question "Do you see what I see?" That most basic of queries binds human beings socially. . . . Having one's perceptions go uncorroborated can make one feel peculiarly alone in the world. . . . marooned on my own private island of navy blue c's, dark brown d's, sparkling green 7's and wine-colored v's. What else did I see differently from the rest of the world? I wondered. What did the rest of the world see that I didn't? It occurred to me that maybe every person in the world had some little oddity of perception they weren't aware of that put them on a private island, mysteriously separated from others. I suddenly had the dizzying feeling that there might be as many of these private islands as there were people in the world.[13]

Synesthesia offers a startling insight: Lower-level brain modules can profoundly affect not only our ordinary sensory perceptions but also how we experience abstract symbols such as letters and numbers. If thought is the manipulation of words and symbols, we need to consider whether our very building blocks of thought might also be subject to involuntary, even genetic, influences that make each of us "private islands" of perception and thinking.

7

When Does a Thought Begin?

Timing, or the Chicken and the Newly Hatched Idea

You mull over an idea; you contemplate, ruminate, meditate, and sleep on it. You gradually are convinced and say to yourself, "Yes, that's right." This apparent cause-and-effect temporal sequence— first the thought, then the assessment of the thought, and then the feeling of correctness—is what gives the *feeling of knowing* its authority. Any other sequence wouldn't make sense and would strip the *feeling of knowing* of any practical value. But experience tells us that the *feeling of knowing* has a variable temporal relationship to conscious "reasoning."

Possible timing sequences could include the following examples. In scenario A, we experience a *feeling of knowing* without any accompanying thought, as is seen with mystical experiences and brain stimulation studies. Any interpretation or explanation of this feeling occurs after the experience. A common contemporary example is a profound spiritual "sense of oneness" followed by the interpretation that this "moment" represented a divine revelation.

In scenario B, a series of unconscious associations is infused with a *sense of correctness*. The thought and the *feeling of correctness* reach consciousness as a unit and are experienced together as an insight or an *aha* moment. Many great scientists have described their breakthroughs as "brainstorms," or "it just popped into my head," rather than as the product of methodical deliberation. They talk of preparation—laying the groundwork—but the actual insight feels like a bolt from the blue. Srinivasa Ramanujan, the famous Indian mathematician, once said that he would "simply know" that a complex result in number theory was true, and that it was only a matter of later proving it.

It is highly unlikely that difficult mathematical theorems can appear without any prior contemplation and preparation. But it is easy to accept that an insight occurs as the result of a new association arising out of the reworking of unresolved prior ruminations, half-formed queries, or vague hunches. These associations begin within the hidden layer and, once judged to be correct, are then passed on into consciousness. We experience the thought and the feeling of its correctness simultaneously as a *eureka* or a *moment of truth*.

In scenario C, an idea is encountered for the first time. It is objectively determined to be correct, and then one "knows" the answer is correct. For example, you know you have found your friend's house when your friend answers the doorbell, or you dial a telephone number and reach the intended party. With scenario C, the feeling of the correctness of a thought clearly follows conscious assessment and testing.

To have unconditional trust that a *feeling of knowing* represents a justifiable conclusion, we need to know which of these three scenarios has actually occurred. Timing is everything. But what if the brain contains mechanisms that rearrange the perception of a sequence of events? What if our brains can trick us into believing that event X follows event Y, even though it actually precedes it? Sounds like a preposterous proposition, but what if this rearrangement is necessary to overcome yet other physiological barriers to the proper perception of a sequence of events?

Optical illusions, once explained, feel like insights into how our brain assembles what we refer to as reality. But when was the last time you saw time presented as an optical illusion?

Subjective Backward Projection of Time

Sandy Koufax's fastball was so fast, some batters would start to swing as he was on his way to the mound.

—Jim Murray

J. Blow sits in his overheated and underdecorated Pittsburgh hotel room deep in a batting slump. His team is in last place; he's a thousand miles from home and a handful of strikeouts from being sent down to a bottom-of-the-barrel farm club. The morning of the game, he gets a call from his wife asking if there'll be a year-end bonus—both daughters need orthodonture and ballet lessons. His six-year-old son comes on and says that he misses his father, then says in a tiny voice that could break a stone's heart, "Please hit a home run for me."

He hangs up and watches a Giants game; Barry Bonds hits three for three, including a double and a home run. Blow opens his wallet and pulls out some crumpled and yellowed scraps of paper—newspaper clippings of quotes by Bonds, Ted Williams, and Stan Musial, three of the greatest hitters in the history of baseball.

Bonds told broadcaster and former Cy Young Award winner Rick Sutcliffe that he had reduced the strike zone to a tiny hitting area, and that's all he looked at. "It's about the size of a quarter," said Bonds.

In 1986, Ted Williams said, "Until I got to two strikes, I looked for one pitch in one area, about the size of a silver dollar."[1]

Stan Musial told a rookie: "If I want to hit a grounder, I hit the top third of the ball. If I want to hit a line drive, I hit the middle third. If I want to hit a fly ball, I hit the bottom third."[2]

How do they do it? Blow asks himself. These days, I can barely see the ball leaving the pitcher's hand.

At the park, the manager is a combination of reassurance and subtle threat. "Just get some wood on the ball. No swinging for the fence. And don't worry. Maybe you can get back your confidence in Springfield [the triple A farm club]. . . ."

Think of all the inputs, conscious and unconscious, swirling through Blow's head as he steps up to the plate. His father is scowling, doubtful, disappointed; his mother is rolling and unrolling the hem of her skirt, softly praying. His high school coach is calling his name; it is the first time that he will be in the starting lineup and he is both cocky and scared. The hidden layer has its work cut out for it; the assignment is no less than weighting childhood slights, long-forgotten failures, prior unexpected triumphs, parental attitudes, and a host of other variables huge enough to sink the psychoanalytic *Titanic*.

Do as the manager says, Blow concludes. You can always explain to your son that winning is more important than personal statistics.

Just connect with the ball. Take a nice easy swing. Blow plants his feet and gets ready.

Here's the windup, and the pitch. . . . A medium speed, no stuff ball, big as a harvest moon, virtually floating toward the plate, Blow thinks, judging from the pitcher's hand at the moment of release. The subcortical motor centers start to salivate. No way that they're going to let this one go. And Blow, fully intent upon just meeting the ball, swings with everything he's got.

He hits a towering home run to left field. The team wins 1–0. Blow is the hero du jour. After the game, the coach asks Blow why he ignored his instructions. Blow says, with complete cortical honesty, "I don't know. Something must have come over me."

PROFESSIONAL BASEBALL PITCHERS throw with velocities in the range of 80 to 100 miles per hour. Elapsed time from the moment of release to the ball crossing home plate ranges from approximately .380 to .460 milliseconds. Minimum reaction time—from the instant the image of the ball's release reaches the retina to the initiation of the swing—is approximately 200 milliseconds.[3] The swing takes another 160 to 190 milliseconds. The combination of reaction and swing time approximately equals the time it takes for a fastball to travel from the pitcher's mound to home plate.[4]

To get an appreciation for the magnitude of the problem, consider that a fastball will travel about nine feet before your retina transmits and your brain processes the initial notification of the ball leaving the pitcher's hand.[5] Full perception of the pitch takes considerably longer. The delay in processing means that when the ball appears to be at a certain position, it is no longer at that position. To see it "where it will be," the brain must integrate the speed of motion over time, estimate the degree of position shift, and combine this with the appearance of the object as seen at the present time.[6] Pretty amazing—a micro-version of precognition, only on a probabilistic level—the "now" that the batter experiences when he initiates his swing is "virtual," generated by complicated subliminal computations.

Once the ball is in flight, it is too late for detailed deliberation. The batter sees the release and the beginning of its path, and then goes on automatic pilot. Sounds suspiciously like an inner machine at the helm, some robotic neuronal clumps that are responsible for a hitter like Babe Ruth or Barry Bonds. Yet we all know that a hitter's skill, beyond mere athleticism, is dependent upon prior practice and extensive study of the game. Great hitters keep extensive notes on the tendencies of opposing pitchers, including what type of pitch and where it will be thrown in various conditions. A 3–0 pitch with the bases loaded is more likely to be down the middle than a 0–2 pitch with the bases empty. The combination of circumstances is infinite, yet each hitter develops a probabilistic profile of the speed, trajectory, and location of the next pitch. It is in this realm that great players have a greater accuracy than novice players.[7]

The act of hitting the ball involves two fundamentally different strategies inextricably linked together—conscious analysis prior to the event, and reliance upon nearly instantaneous subconscious calculations at the onset of the event. The cortex sets out general guidelines for when to swing and where, then hands the controls over to quicker subcortical mechanisms.[8] A simplified schema provided by a computer scientist-engineer after extensive study of the physics of a pitch: "We divide the pitch into thirds. During the first third the batter gathers sensory data; during the middle third he does computations (predicting where and when the ball will collide with his bat); during the last third he is swinging. *During the swing he could close his eyes and it would not make any difference.* He can't alter the swing. The most he can do is check the swing."⁹ (Italics mine.)

These studies have been duplicated with a variety of other sports, from Ping-Pong to squash and cricket. Can you imagine a boxing match in which a fighter waits until he has fully seen and analyzed a punch before deciding what to do? The survival benefits of immediate action are self-evident.

In conclusion, the batter is swinging at previously determined probabilities, not a closely observed ball. A fabulous hitter such as Barry Bonds is better at fine-tuning his swing in midarc than your average batter, but this does not result from conscious perception, deliberation, and then a decision. There simply isn't enough time.¹⁰ Yet Bonds, Williams, and Musial swear that they can gauge their swing to within a baseball diameter target, or less. A truly extraordinary feat when even the most advanced physics applied to the ball's early flight path cannot make such a precise prediction.

So, are the world's greatest hitters really wishful thinkers—I hit the ball; therefore I saw it as it approached the plate? How can we balance off what the players believe that they saw with what science tells us is physiologically possible?

"Now" You See It, "Now" You Don't

When we look out at the universe, it is easy to understand that the light from the sun takes nine minutes to reach the earth and that we are looking at a nine-minute-old event. Ditto for the light-years for light to reach us from a distant galaxy. We have no difficulty living within a nonsimultaneous universe, with both present and past being represented as *now* on our retinas. The distances are simply too great to make a difference in our daily lives. But what about a rapidly approaching baseball?

Hitting coaches stress "keeping your eye on the ball." Some say that you can see the ball to within a few feet of the plate; others believe you can see the ball strike the bat. No matter; what is peculiar is that such images would not reach consciousness until after the swing has been made and the ball is already on its way out of the park or in the catcher's mitt. If the brain did not somehow compensate and project the image of the approaching baseball backward in time, you would see the ball approach the plate after you had already hit it.

This aberration in the fabric of perceived time has been hotly argued as representing everything from evidence for noncausality to intention preceding awareness. But the explanation needn't be deeply philosophical. This coordination of inputs is an everyday occurrence. If you bump into a door, the sensory inputs from your nose reach the brain sooner than those from your big toe, yet you perceive hitting the door with your entire body all at once.[11] The brain adjusts for these time lags. When I tap my foot, the motor movements are felt to be synchronous with my foot striking the ground. The length of time that it takes the sensation of my foot hitting the ground to reach the brain and be processed is not apparent. Without such adjustments, the varying delay between sensory inputs would create a kaleidoscopic sense of time, a present that is spread out over time (a "thick" present), as opposed to an instantaneous "now."

Color Phi

If you want to see subjective backward projection in time, try a simple experiment.[12] When lights in close proximity are briefly lit in rapid succession, we will see a single light moving from point A to point B (the basis of apparent movement in old-fashioned sign marquees). The brain interprets these two flashes of light as though the light is moving between the two points.

Now color the lights. Make point A red and point B green. As we see the light move from point A to point B, it will abruptly shift from red to green approximately at the midpoint between the two lights.[13] In other words, we will see the green light flash prior to it actually being turned on. Within the interval provided by retinal-cortical transmission and processing, by mechanisms still unknown, the brain has shoved the flashing green image backward in time (we experience it sooner). By taking advantage of the window of time required for processing incoming sensory data before outputting it as perception, discordant brain time and "external" time are re-aligned to allow for perception to create a seamless world of "now." It has been estimated that the brain routinely can smooth out the discrepancies by backward projection of the second image by as much as 120 milliseconds.[14] According to this bizarre but necessary neurophysiology, "being in the moment" is a virtual recipe that steals from both the recent past and the immediate future.

To further complicate the problem of the timing of perception, consider how different the approaching baseball looks to the batter and to you, an observer sitting behind home plate. The pitcher fires three successive ninety-five-miles-per-hour blazers. The batter whiffs the first and fouls off the next two. He prepares himself for another smoker. Instead, the pitcher lobs a deceptive sixty-five-miles-per-hour change-up. The batter swings far too early and strikes out. You watch in amusement and ask yourself how the batter can make five million a year and so misjudge a ball that, to your uninvolved eye, a Little Leaguer could hit.

The difference is that while the batter's decision to swing begins prior to his full conscious appreciation that the pitcher has thrown a slow pitch, you have the luxury to see the ball's entire path toward the plate. By not being forced to immediately decide whether or not to swing, you see a batter being badly fooled by a pitch that doesn't fool you.

The basic neurobiological principle is that the need for an immediate response time reduces the accuracy of perception of incoming information. Though most of us aren't involved in high-speed sports, we all experience these limitations in the most crucial of daily activities—normal conversation. Indeed, conversation is as much a high-speed competition as a top-flight table tennis match.[15] First consider the act of listening. We are bombarded with the rapid presentation of individual phonemes strung together to make words, phrases, and sentences. Processing takes time. A word may not be initially decipherable; only with further speech is it clarified. Think of how we listen to someone with a foreign accent or regional dialect. We hold a phrase in short-term memory until it is put into context. Watch a modern speech recognition program in action and you will see words being corrected as more information (further words) is inputted.

For example, in testing a new speech recognition program, I dictated the phrase, "No cuts, bruises, or lacerations." The program typed out, "No cuts, bruises, or lesser Asians." I tried to speak as slowly and distinctly as possible, but without success. It wasn't until I added the phrase, "The patient's X-ray showed a hairline fracture," that the program, after a pause of several seconds, corrected *lesser Asian* to *laceration*. The program needed more information to improve its accuracy.

Our recognition of speech works in a similar manner. Over

time, we build up massive neural networks that recognize letters, words, phrases, personal syntax, and so forth. Try dictating, "He's a wolf in cheap clothing." The speech recognition program, if possessing the original phrase in its database, will keep typing "sheep's clothing." Unlike a computer-driven speech recognition program, we have the added benefit of seeing body language and gestures—all the nonverbal clues that give additional hints at meaning. By judging the speaker's delivery, the presence or absence of a smile or blank expression, we are better able to determine whether the choice of words was intentional (a pun) or unintentional (a malapropism). This interpretation might take considerable additional time, after which we can correct our original impression. The pleasure of the unexpected punch line or the misunderstood homonym underscores how meaning is contextual, and contingent upon what hasn't yet been said.

Now visualize conversation as a means for the exchange of complex ideas with each participant's response dependent upon whether or not he believes the idea is correct. Instead of throwing a fastball, each discussant is throwing an idea at the other. If the listener judges the idea as correct, he will not swing (he will accept the idea as is). If he thinks the approaching idea is incorrect, he will swing (formulate an immediate rebuttal and/or interrupt the speaker to interject his correction).

Here's the windup, and here's the thought. The listener's decision as to the thought's correctness will be based upon a quick glimpse of the idea leaving the other's lips, snap judgments of body language, sighs, gestures, facial expressions, and all the various verbal and nonverbal contributions to interpretation of the spoken word. If the listener is forced to make a quick response, the decision as to the thought's correctness will be subject to the same physiological restraints as a batter's assessment of an incoming pitch. Nevertheless, due to the subjective backward referral of time, the listener will feel that he fully considered the idea *before* deciding on its correctness (the equivalent of Barry Bonds believing that he can see the ball in a quarter-sized strike zone *before* initiating his swing). A great baseball player bats .300; a .300 conversationalist is strictly minor league.

How different conversation sounds when we don't feel obliged to respond. As uninvolved spectators luxuriating in our more leisurely processing time, we easily see the shallowness, evasiveness, and lack of real exchange of ideas in most dialogue. We know better than to trust most on-the-fly conversations; we gripe about sound bites and MTV emphasis on quickness of response over dead air silences. We cringe at the obvious nonresponsive answers that characterize presidential candidate debates. But nothing changes. Sadly, the problem is at least in part a matter of the physiology of conversation. As we move from silent observer to active discussant, we become mired in the very processing problem we're trying to overcome. Given the time constraints of rapid-fire conversation, the *feeling of knowing* will be triggered prior to full perception of the incoming idea, yet feel as though it followed consideration of the idea.

AT THE SHORT end of the temporal spectrum it is possible to see how such subjective backward referral of the *feeling of knowing* might lead to erroneous conclusions, but temporal illusions also occur over a much longer time span. Which brings us to a critical question: When does a thought begin? With the baseball example, we can detect an altered perception of the sequence of events because we can measure the speed of the ball as well as the conduction velocities of electrical impulses within the central and peripheral nervous system. But how are we to measure the timing of a thought?

The *feeling of knowing* can follow a thought: "What is Ima Klutz's phone number?" (Scenario C from the beginning of this chapter.) You check the telephone directory and find five identical names and numbers. You try the first, not knowing whether or not it is correct. When Ima answers, you immediately *know* that the number is correct. The *feeling of knowing* follows hearing Ima's voice on the phone.

But once we leave the simplest cause-and-effect examples, we are on thin ice. The emergence of a complex thought involving new associations can vary from milliseconds to decades. I might pass a woman in the street today, and suddenly tomorrow (or so it seems) remember a girlfriend from long ago. The time for germination of an idea for a new book can be years. Until a thought reaches awareness, it is inaccessible to standard scientific measurements—a silent traveler invisible within the hidden layer. But we can try some simple thought experiments to see if we can draw any conclusions.

IZZY NUTZ LIVES at 123 Filbert Street. You have been invited for dinner, but have never been to his house. You are driving along Filbert Street when you see the signpost for 123. In this example, we have a pretty good idea of when we thought, "That's 123 Filbert," and when we knew that the thought was correct. You perceive the 123 sign, and then say, "That's it."

Now consider an alternative tale. You've been to Izzy's house twenty years ago with your wife, and think that you remember it quite clearly. This time, it is dark and stormy; the street signs in the neighborhood have been blown down. "No problem," you tell your wife, "I remember his house like the back of my hand." (Anyone who's ever traveled with a spouse knows how this story will turn out.) After much bickering, you pull onto a street that looks just like the Filbert Street of your memory. "Trust me," you say to your wife, who is contemplating the single life. You see a house that looks exactly like Izzy's. "There it is."

"Are you sure?" your wife asks. "It's not at all like I remember it."

"Yes. I know this is the house."

You get out, ring the doorbell, and are told by the occupant that this is not even Filbert Street; "Izzy lives one block over." Back in the car, your wife shrugs in disgust while you try to shake the strange idea that the man inside the house is wrong. It must be Izzy's; it's just as you remember it. "I guess I was wrong," you admit reluctantly, and then add, "I could have sworn it was his house," still not entirely convinced.

In this case, when did you "know" that it was Izzy's house? Twenty years ago, you and Izzy spent the evening in his living room. At that time you had overwhelming evidence that the house was Izzy's. Being essential to the learning process, the *feeling of correctness* merged with the memory of the evening to form the neural network that represents Izzy's home. Imagine this network functioning like an old telephone switchboard in which several circuits are connected as in a party line. There is not a bit of privacy—the circuits remain in constant communication—everyone hears everyone else. The image of the house and *knowing* cannot get away from each other. Twenty years later, both are activated when you see the house that is similar to the stored image of Izzy's place.

The felt sequence is that you see the house and then say to yourself, "Yes, that one is Izzy's." No other timing would make sense. Imagine how confused you'd be if the feeling occurred before you felt that you saw the house. And yet it was the twenty-year-old *feeling of knowing* that allowed you to recognize the house that you then said must be Izzy's.

Despite being an everyday event, temporal reordering remains poorly understood. Without implying that we have a clue as to the underlying anatomy or physiology, conceptually there must exist a central time synchronization mechanism. Though this cumbersome mouthful tells us nothing about specific brain activity, it does help us to recognize that our internal "brain time" may not be an accurate reflection of "external time," and that the brain is capable of smoothing out internal-external time discrepancies to suit its own purposes.

For those thoughts that activate prior thoughts and memories, we cannot know what portion of thought is presently being formed, what is being remembered, or when the *feeling of knowing* occurred. What might seem like cause-and-effect—A before B and causing C—cannot always be trusted to be the correct sequence of events. Brain time has its own agenda.

8

Perceptual Thoughts: A Further Clarification

YOUR ERRONEOUS RECOLLECTION OF IZZY NUTZ'S HOUSE also raises the knotty problem of the reliability of memory. If we see the 123 Filbert Street sign, and Izzy answers the door, the *feeling of knowing* is appropriate. But what about the second scenario, when it wasn't Izzy's house? The *feeling of knowing* was the same, only this time it was not to be trusted. To understand the problem of the same *feeling of knowing* being attached to correct and incorrect conclusions, we need to take a quick look at the present-day understanding of memory.

Episodic Versus Semantic Memory

My grammar school was torn down decades ago; my high school has been converted to a public administration building. But the names remain engraved in my brain. Neuropsychologists refer to these as *semantic* memories, in contrast to *episodic* memories, which are the memories of what happened at the schools.

Semantic memories include everything from the date and time of the attack on Pearl Harbor and the number of home runs hit by Babe Ruth to your present address and social security number. These are the packets of concrete information that can be externally verified and agreed upon. We can count the number of floors of the Empire State Building. I can pull out my old high school yearbooks and see the name Lowell High School embossed on the cover. A foot will always be twelve inches.

In contrast, *episodic* refers to the remembrance of specific episodes strung together via a narrative of *first this happened, and then that happened.* These are the memories that are revised by subsequent experience.[1]

"I Witness" Accuracy

Spend time reminiscing with a sibling and chances are that you will uncover dissimilar accounts of what you thought were shared childhoods. My sister and I might as well have been raised on separate planets, there is so little overlap in our respective tales from the crib. Even the rubbery Sunday chicken was fresh or frozen, bland or spicy, served hot or cold. I have a friend whose sister published her memoir of her childhood; while reading it, my friend kept checking the jacket photo to be sure that it was his sister who had written the book.

Let us assume that my sister is right, that the chicken was exactly as she now describes it, and that I originally saw exactly what she saw. (This is not a concession, only a hypothetical.) Now, my memory is different than hers. But I am not equipped with an alarm mechanism or pop-up dialogue box that warns me when a memory is altered. I was not notified; I never experienced the mutating of a former memory (I use the word *mutate* intentionally). If such changes occur in an untraceable silence, I must concede that I am the quintessential postmodern unreliable narrator. The *I* of prior experience is only a fleeting pattern of no particular predictability; I am nothing if not my past.

None of us have an instinctual belief that our memories are this fragile. Despite the proliferation of psychological studies questioning the accuracy of episodic memory recall, we cling to the belief that our pasts approximately correspond to our memories. Sometimes we sense that the details are a bit faded, but we rarely doubt the essence of a memory. We rely upon the notion that, at the very least, the memories of our past reflect fundamental truths.

The all-too-seductive argument: If I can be sure of where I was born, and this *feeling of knowing* can be easily verified, shouldn't I trust all my memories that feel correct? If I can sing the complete lyrics of an obscure Beatles song and double-check its accuracy on some Internet Web site, then surely I can remember two lines of dialogue from that dreadful conversation during which you accused me of . . . or promised me . . . or "I specifically told you that. . . ."

Dialogue is dialogue. Memory is memory. Right? If you think the brain regularly accomplishes this, then you've never had the following exasperating exchange.

"You started it."

"No, first you said, and then I said . . ."

"Just once, why can't you get it straight? You said, and then I said . . . , and, by the way, that's not at all what I actually said."

"I heard what I heard. You started this one by accusing me of . . ."

"I hadn't even opened my mouth yet. Not a peep."

"So now it's my imagination? We should videotape our conversations."

"We couldn't even agree on when to start the camera."

If we are so easily confused over who said what to whom when, how can we consider these memories accurate? Yet that's how we live our lives. If you have any question about episodic memories' inherent instability and unreliability, you need only consider the *Challenger* study, UFO and alien abduction testimonials, or the O. J. Simpson trial. (This is not the place to cite chapter and verse about the frailties of episodic memory; for an excellent summary of the latest studies of faulty recall, shifting memories, and false memory syndromes, check out the excellent and easily accessible writings of Harvard psychologist Daniel Schacter.)[2]

If we accept that there are two fairly distinct forms of memory—semantic and episodic—we might also contemplate the possibility of an analogous distinction for different categories of thought. At one end of the thought spectrum would be brute memorization and rote utilization of facts as tools. If, in high school physics, you learn that $f = ma$, you have memorized an equation that won't change with subsequent experience. If a quantum mechanic should crawl out from under an atom and say that $f = ma$ doesn't work, your memory of the equation remains correct. Memorizing facts doesn't require logic, cause-and-effect, or any significant ability to reason.

Some thoughts, like semantic memories, are essentially self-defining—Christmas is on December 25. A foot is always twelve inches. No complex hidden layer processing is necessary; a fact will remain a fact in perpetuity (as long as the initial underlying assumptions persist). It isn't necessary to reason out that Christmas is on December 25, because December 25 is part of the definition of Christmas.

For purposes of simplicity, let's call thoughts that require only memorization, but no decision making, logical analysis, or reasoning, *semantic thoughts.* Not a very appealing term, but useful as a reminder of the similarity to *semantic memories.* By contrast, thoughts that arise out of complex computations within the hidden layer might be seen as the equivalent of episodic memories that are continuously and subliminally undergoing revisions, augmentations, and diminutions. Like episodic memories, such thoughts require an element of perception and are subject to a variety of perceptual illusions. Since the term *episodic thought* is cumbersome, I have chosen the more descriptive term *perceptual thought.*[3] In the following discussion of thinking, we will be primarily addressing perceptual thoughts.

9

The Pleasure of Your Thoughts

IT IS OBVIOUS THAT THE *FEELING OF KNOWING* IS ESSENTIAL to the learning process, but to appreciate its enormous power requires a brief discussion of brain reward systems.

I AM AN inveterate poker player. To justify my degeneracy, I have been known to mumble about the thrill of competition; a fascination with the quick calculation of the proper odds; a level playing field (taking steroids doesn't help); the best hand *always* wins (there are no bad line calls, flexible strike zones, or hanging chads). To lift a title from Paul Auster, I may even wax poetic on the music of chance in a world of unpredictability.

All the above might be true, yet I must confess to a more powerful motivation: I play to feel lucky.

"Not me," protests the statistician. Poker isn't gambling; in the long run, the cards will break even and skill will prevail. To be fair, I have met players who outwardly pay 100 percent lip service to the laws of probabilities, but I suspect this is merely a great poker face. At the moment the crucial card is being dealt, you show me a gambler, no matter how icy his cerebrospinal fluid, and I will show you primitive man howling at the moon, waiting for the miracle that will deliver him from ordinariness. Stick the most rational of the rationalists in a poker game, hook up a lie detector to his subconscious, and you will hear the silent supplications. Oh, poker lord, give me an ace.

Check out the raw expressions of hope in those gathered around the roulette wheel, the 7-Eleven store at lottery time, or those transfixed by the NASDAQ ticker in Times Square. Stand in the Las Vegas baggage claim arrival area and you are knee-deep in electricity and excitement. Those impatiently waiting to grab their bags and hit the tables ignore the body language of defeat and the expressions of tired truths worn by those leaving town. The dismal likelihood of "the big win" is momentarily ignored. According to the tortured odds of wishful thinking, the knowledge that nearly everyone else has lost only means that your chances of winning must be greater. ("Let's play that slot machine. It hasn't paid off in days, so it must be due.")

The recent stock market bubble was in large part dependent upon an irrational suspension of disbelief. People talk of investing, but the thrill of watching one's stocks go up wasn't just about making money. Despite the parabolic rise in stock valuations, the vast majority of analysts discarded the lessons of history in favor of the second-by-second intoxication of an out-of-control market. We all should have known better, but we could not help ourselves.

The Pleasure Principle

If you place brain electrodes in strategic rat pleasure centers, the rats will continuously press the bar that activates the plea-sure-producing electrodes, forgoing food and water, until they drop.[1] Through the use of imaging and detailed anatomic stud-ies, as well as microelectrode implantations, neuroscientists have demonstrated extensive connections between the regions of the brain responsible for pleasure-reward systems, affect and emo-tion, and opioid peptides (endorphins). A key component of brain reward circuitry is the *mesolimbic dopamine system*, a set of nerve cells that originate in the upper brain stem (the ventral tegmental area). Though several neurotransmitters may be in-volved, dopamine is considered essential to the activation of this reward circuitry.[2] This system extends to those areas that inte-grate emotion and cognition, including portions of the limbic sys-tem and orbito-frontal cortex, and the nucleus accumbens—an area at the base of the brain widely thought to be involved in ad-dictive behavior.

On brain imaging studies, we can see naturally occurring re-ward systems hard at play—clumps of neurons positively radiant in response to pleasing tastes, odors, touch, and music.[3] Man, in his greater psychopharmaceutical ingenuity, has connived ways of tricking the brain—cocaine, amphetamines, alcohol, and nicotine activate similar regions.[4] Winning at gambling turns the orbito-frontal cortex into pure neon.[5] Without this exhilaration, there would be no addiction. Conversely, obliteration of the orbito-frontal region via prefrontal leucotomies (lobotomies) results in apathetic and unmotivated human zombies; long-term intent is abolished.[6]

Research into addictions to drugs, alcohol, gambling, and cigarettes have been instrumental in revealing how behavior is rewarded. The general principle equally applicable to the worst cocaine addiction, stamp collecting, or idle musings is that for a behavior to persist, there must be some brain-mediated reward.

The question we now need to address: What is thought's reward system?

IF THROUGH SNAP judgment or insightful deliberation, you avoid a charging, hungry lion by scurrying up a tree, you have concrete evidence of the value of your thoughts. The lion slinks away and settles on gazelle tartare for lunch. You climb down from the tree feeling that you have learned something. The *feeling of knowing* and the decision to climb the tree become linked together in the neural network labeled "what to do in the case of a charging lion." The more powerful the experience and the more times it occurs, the greater becomes the linkage between the decision and the feeling that the decision is correct.

The *feeling of knowing* and the related feelings of familiarity are as integral to learning as the visual system is to seeing, the olfactory system is to smell, as basic as mechanisms for fight or flight.[7] Feelings of strangeness and unfamiliarity can warn us that we are making a wrong turn in our thinking. ("That doesn't feel right." "Something's rotten in Denmark." "No way; bad vibes all around.")

The *feeling of knowing* most likely was thought's initial yes-man. "You are one smart dude," the feeling exclaims, high-fiving you, often followed by further self-inflation such as "That's using your head." Man has evolved. Thoughts have become more complex and abstract; much of what we think about today has no clear answer, no obvious cause-and-effect result, and isn't easily measurable. We can never know with certainty whether decisions to invade Iraq, to restrict stem cell research, or to permit private ownership of handguns are the best decisions. The law of unforeseen consequences tells us that today's seemingly positive result might be next decade's catastrophe. (Remember DDT?) Personal decisions, from deciding whether to get tested for the genes for Alzheimer's to whether or not to title your novel *Catch-22* cannot be tested. So much of our thinking occurs in the dark.

Our catch-22: In order to pursue a new thought, we must feel the thought is worth pursuing *before* we have any supporting evidence or justification. Otherwise, we would only consider ideas we already know to be correct. But what would the reward be for a new or unique idea? We talk of the pleasure of knowledge for knowledge's sake, but this presumes that what you are acquiring is bona fide knowledge. Proceeding without any sense of a thought's value isn't a high-priority activity. Just watch your kid avoid his homework, complaining bitterly that studying Latin or logic is worthless. "What's the point?" is nothing more than thought's reward system switched to off (or running on empty if you prefer the neurochemical metaphor).

I Can't Go On, I Must Go On

I hate crossword puzzles, but have plenty of friends who are addicted. Twenty-six across: a six-letter word for intoxicated. Begins in *s*. Ends in *d*. You run through all the words that you can think of, then systematically test them. Stewed? Stoned? After a period of struggle, you come up with a word that fits with the remainder of the puzzle—soused. You are rewarded with a little frisson of "I figured it out."

You do this a few times and you are hooked.

With this crossword puzzle example, feedback is fairly immediate. Once a few words are in place, you can quickly assess further choices. Now expand the scope of the challenge. Imagine single-handedly tackling a football-field-sized puzzle. Years will pass before any pattern is discernible. Until then you cannot backtrack to see if your choices fit with other parts of the puzzle. Most of us would throw up our hands in defeat.

Unless there's a major reason to finish. What if the puzzle is the key to escaping from a life sentence in a nightmarish Third World jail that makes Abu Ghraib look like the Plaza? Your life depends upon finishing the puzzle as accurately and quickly as possible.

When you begin, each slot has so many possibilities. You cannot imagine trying to match 25999000 across with 45999990 down without some connecting words. You long for a crumb of encouragement, a nice warm mental pat on the back. Without evidence, you are willing to settle for false hope and irrational alternatives. You search your heart. If you believe in divine revelation, you can have God's personal guarantee that the word selection *must* be right. Your choices, blessed with a highest authority's seal of approval, are unassailable. But what if you lack this sense of faith? Is the solitary pleasure of unsubstantiated thoughts enough? Remember, this project is going to take years, not MTV milliseconds.

Most physiological reward systems are measured with a stopwatch, not a calendar. With fight or flight, you know pronto whether running away was the right choice. Cocaine and gambling are *now* rewards. No one ever listened to Bach with the goal of experiencing enjoyment in a month, or told a joke to make you laugh next year. Pleasure systems don't have a memory; they're now or never, measured within the time frame of synaptic transmission and neurotransmitter metabolism. Continued reward requires continued brain stimulation. Even rats have figured this out.

Present-day fMRI studies of reward systems measure short-term outcomes. A volunteer plays a video game while nestled inside an MRI scanner; responses to winning or losing are imaged during a single scanning session. Longer-term studies are fraught with interpretative minefields as well as enormous and perhaps insurmountable logistical challenges. For the foreseeable future, whether investigating the reasons for finishing the stadium-sized puzzle or devoting a decade to obsessive ruminations on an epic poem, our understanding of long-term reward systems will be unprovable extrapolations of brief-duration studies.

A further problem is that there are myriad studies on reward systems and various aspects of thought related to specific subjects—such as fMRI studies on the pleasure centers involved in thinking about winning, sex, drugs, and so on. But it is much harder to construct a study to examine how we think about the very process of thinking, and how we reward ourselves for wide-ranging hard-to-categorize ruminations. Imagine asking a volunteer jammed inside an MRI tube to indicate whenever he was deep in daydreaming or philosophical musings. The mere requirement to signal when a particular category of thought was occurring would alter both his baseline and activation fMRI images (a vivid demonstration of Heisenberg's contribution to behavioral neurology).

I CONFESS TO a real discomfort with explaining all human actions exclusively in evolutionary terms. Just as so much of Freud's phallic presumptions have turned out poorly, today's reliance upon adaptive explanations may also be too simplistic. Using behavioral observations to determine why a physical characteristic evolved might lead us to conclude that the human appendix developed as a source of mortgage payments for hungry surgeons. There is nothing wrong with the idea that our biology evolved and is adaptive; rather the problem is in knowing exactly what that adaptation must have been. Today's commonsense transparencies might be tomorrow's historical jokes. Nevertheless . . .

The Big What-if

When our brains stumbled across the potential for abstract thought, an appropriate reward system was necessary. Though a wide variety of pleasures could offer short-term solutions— appreciation of the intricacies of a thought, the effort involved, the beauty of a particular sequence of numbers, the elegance of syntax, or the romantic ideal of devoting oneself to just thinking— none seem sufficiently powerful and long-lasting to carry our thinking forward through long nights of doubt and despair. Without some demonstration of a practical value of a thought, it would seem pointless to persist indefinitely.

If you doubt the need for a reward system for the unprovable thought, stop and consider what propels you forward in a long-term intellectual project. In writing a novel, there are the dark days when you want to rip up the manuscript and swim laps in Jack Daniels. You hound your wife, chasing her around the kitchen reading your favorite lines, pleading for praise. You phone up a friend, read him a passage, knowing that he will tell you it's okay even if it isn't. You are buffeted by the dry wind of meaningless pursuit.

Sooner or later, you need some personal conviction that your novel is working. In matters artistic, you rely upon a sense of aesthetics, a feeling that you have captured an essential inner truth or vision. Whatever you call the feeling, it contains a connotation of value, of achievement and direction. In matters scientific, we look for a feeling that our ideas, though presently incomplete, do represent reasonable building blocks that might one day coalesce into an established fact or theory.

A wonderful glimpse of the relationship between pursuit of an idea and a sense of rightness as both reward and motivation is provided in a biography of the renowned physicist Erwin Schrödinger.

> As a student in Vienna, Schrödinger was devoted to mathematics, to poetry and to nature. It seems characteristic of his generation of scientists that they were not afraid to admit that an aesthetic impulse moved them, that they were chasing a glimpse, however fleeting, of some confirming, self-ratifying idea of beauty, an equation to transcend all equations: *some sense of perfect rightness, a feeling of the universe clicking into place.*[8] (Italics mine.)

Perhaps you disagree as to the nature of the reward. A desire to succeed, a burning ambition, a need for promotion, an I'll-show-you attitude; whatever psychological motivations that you assign to a behavior do not address the underlying physiology of how the brain rewards such behavior. No matter what the psychological impulse, no one ever spent twenty years in some stinky lab without some little pellet of pleasure periodically dropped onto his median forebrain bundle.

The choices are to either develop a new reward system specific for this emerging ability of thought, or to expand the role of existing systems. Economy of effort would favor the latter. The *feeling of knowing* was already securely in place as a feedback reward system for learning. What if the feeling could be repackaged as a motivation for pursuing the unproven thought?

A perverse possibility: An unwarranted *feeling of knowing* might serve a positive evolutionary role.

The notion of the empiric method is based upon the simple premise of trial and error. An erroneous initial idea that prompts further investigation is preferable to no incentive to any thought at all. With crossword puzzles, you don't expect all your first choices to be the final ones. Ditto for working out equations for tough math problems, designing your home, or writing a symphony. The history of science is the history of successive approximations.

The problem is that we need a reward strong enough to tide us over until our thoughts can be verified. And, to be convincing, it must feel similar to the feeling we get when we know a thought is correct and can prove it (as in getting the right phone number).

Enter a spectrum of bridging motivations ranging from hunches and gut feelings to faith, belief, and profound certainty. From vague inklings of familiarity such as déjà vu to an overwhelming sense of conviction, the various sensations that contribute to a *feeling of knowing* have evolved an additional function. Say hello to abstract thought's subliminal cheerleader.

In California, the pedestrian has the right-of-way (at least in theory). Before coming to New York City, I'd crossed San Francisco streets thousands of times, and the cars always stopped. My first day in Manhattan, unaware that the pedestrian's only rights are last rites, I was crossing an intersection with a city-raised college friend. A cab was barreling down the street, aimed directly at us. My friend bolted for cover. I had the same urge, but resisted.

I'd run the experiment thousands of times before and had always come to the same conclusion. The car will stop before hitting me. I stood my presumed legal ground and gave the oncoming cab a defiant staredown. My friend hollered from the safety of the sidewalk, but I refused to listen. I had rights. Instead of braking, the cab accelerated. I jumped back just in time. The cabbie laughed, gave me the finger, and sped on.

Back on the sidewalk, my friend said, "I warned you, but no, you needed to find out for yourself." He added, with a tight Midtown smile, "Indecision is the mother of discontent."

I will never know whether or not the cabbie would have swerved away at the last second. I will never know if my initial decision to stand my ground would have been right or wrong. This is not a question best answered by a controlled study or methodical trial and error.

One of the alleged virtues of a mature person is the ability to delay instant gratification. Stand in line in front of a Good Humor ice cream truck on a hot summer day, while holding the *Harvard Health Letter* warning of the dangers of obesity and cholesterol— then ask yourself which reward systems bring the greatest pleasure. A central conflict of civilization—basic urges versus more level-headed and considered responses—is ultimately a contest between immediate pleasures and longer-term rewards. (A biologically based preference for immediate gratification certainly provides a glib explanation for our stunningly shortsighted attitudes toward foreign policy, ecology, global warming, and population control.)

Double-Edged Single-Mindedness

In order to pursue long-range thoughts, we must derive sufficient reward from a line of reasoning to keep at the idea, yet remain flexible and willing to abandon the idea once there is contrary evidence. But if the process takes time and a repeated sense of reward develops, the neural connections binding the thought with the sensation of being correct will gradually strengthen. Once established, such connections are difficult to undo. Anyone who's played golf knows how difficult it is to get rid of a slice or a hook. The worst part is that the bad swing that creates the slice actually feels more correct than the better swing that would eliminate it. You address the ball with the horrible dilemma of feeling more comfortable with a stance that you know is incorrect. If old patterns were easy to break, so would par. This is particularly so with emotional habits.

Using repetitive electrical stimulation of the amygdala, Joseph LeDoux has produced conditioned fear responses in rats that persisted throughout the rat's lifetime. LeDoux concluded that, once formed, such networks are indelible, and that an "emotional memory may be forever."[9] Similar results have been seen in addiction experiments. Hook rats on cocaine, heroin, amphetamine, and other habit-forming drugs and the animals will self-administer the drugs at the expense of normal activities such as eating or drinking. When the substance is taken away, the drug-seeking behavior is eventually abandoned, but the reward isn't forgotten. A rat that has remained clean for months will quickly return to its drug-seeking behavior when given just a taste of the drug again or even if it is put back in the same environment in which it became addicted. The mere sight of the apparatus that administers the drug is enough to kick in the behavior again.

The studies are impressive; once established, emotional habits and patterns and expectations of behavioral rewards are difficult to fully eradicate. This same argument applies to thoughts. Once firmly established, a neural network that links a thought and the *feeling of correctness* is not easily undone. An idea known to be wrong continues to feel correct. Witness the *Challenger* study student's comment, the geologist who accepts the overwhelming evidence of evolution, yet continues to believe in creationism, or the patient who continues to believe that his sham surgery repaired his knee.

I often wonder if an insistence upon being right might have physiological similarities to other addictions, including possible genetic predispositions?[10] We all know others (never ourselves) who go out of their way to prove a point, seem to derive more pleasure from final answers than ongoing questions, and want definitive one-stop-shopping resolutions to complex social problems and unambiguous endings to movies and novels. In being constantly on the lookout for the last word, they often appear as compelled and driven as the worst of addicts. And perhaps they are. Might the know-it-all personality trait be seen as an addiction to the pleasure of the *feeling of knowing?*

In the early nineties, biochemist Richard Ebstein and colleagues at Hebrew University in Jerusalem asked volunteers to self-rate their desire for risky or novelty-seeking behavior. His finding was that the higher the degree of such behavior, the lower the subjects' levels were of a gene (the DRD4 receptor gene) that regulates dopamine activity in crucial mesolimbic structures.[11] His hypothesis is that people engage in more risky or exciting behavior in order to stimulate a less responsive dopamine-based reward system.

More recently, in studies of subjects reporting a higher degree of selfless or altruistic behavior, he has found higher levels of the same gene, as though greater amounts of the gene allow the same degree of pleasure from less exciting activities than in those who lack the gene. Ebstein has postulated, "This may mean that people who don't get enough dopamine in their brains seek out drugs or other such means to get a 'high.' Dopamine probably plays a key role in pro-social behavior. People with the altruism gene may do good works because they get more of a thrill out of their good works."[12]

As with most questionnaire-based studies, there remain considerable difficulties with interpretation as well as replication. Also, the correlation between socially responsible behavior and dopamine metabolism seems overly simplistic. But what does emerge from these studies is that genes can affect the degree of responsiveness of brain reward systems. It seems highly likely that the same argument can be made for the reward systems for thought.

I cannot help wondering if an educational system that promotes black or white and yes or no answers might be affecting how reward systems develop in our youth. If the fundamental thrust of education is "being correct" rather than acquiring a thoughtful awareness of ambiguities; inconsistencies, and underlying paradoxes, it is easy to see how the brain reward systems might be molded to prefer certainty over open-mindedness. To the extent that doubt is less emphasized, there will be far more risk in asking tough questions. Conversely, we, like rats rewarded for pressing the bar, will stick with the tried-and-true responses.

To extend the reward system-addiction analogy, I also wonder if each of us experiences a different degree of pleasure out of the *feeling of knowing* in the same way that we each respond differently to mind-altering drugs or alcohol. Contrast the following two quotes. Do they represent purely philosophical differences, or are inherent biological predilections playing a role?

I can live with doubt and uncertainty and not knowing. I have approximate answers and possible beliefs and different degrees of certainty about different things ... It doesn't frighten me.

—Nobel laureate Richard Feynman

Dear Mrs. Burton,
Thanks for giving us a tour of the museum. I was the girl who raised my hand all the time and knew all the answers.

—A thank-you note to my wife from a precocious seven-year-old

Judge a man by his questions rather than his answers.

—Voltaire

The *feeling of knowing* is essential for both confirming our thoughts and for motivating those thoughts that either haven't yet or can't be proven. These two roles can be both complementary and contradictory, and can lead to an unavoidable confusion as to what we feel that we know—a confusion that cannot be entirely resolved without taking away the reward system for long-range thoughts. If we are to understand why certainty is such a common state of mind and so difficult to shake, we need to grapple with several fundamental questions.

What are the biological rewards for pure thought and how are they related to the *feeling of knowing*? Are there inherent individual differences in the degree and quality of expression of these rewards, including the potential for addiction? Can these differences be addressed via behavioral changes and shifts in educational emphasis? Can we learn to sense greater pleasure out of feelings of doubt in the way that some people derive more pleasure from questions than answers? Are there ways to adjust such systems to optimize learning and motivate long-range intellectual pursuits without overshooting the mark and promoting dogmatism and an excessive or unjustified *sense of conviction?*

In summary, any present-day understanding of how we know what we know must take into consideration the contradictory nature of thought's reward systems. The *feeling of knowing,* the reward for both proven and unproven thoughts, is learning's best friend, and mental flexibility's worst enemy.

10

Genes and Thought

I OCCASIONALLY ATTEND A BOOK CLUB COMPOSED PRIMA-
rily of University of California professors, software designers, and
venture capitalists. They rarely read novels or poetry, because
"such books do not lend themselves to lively exchanges of ideas.
They're just feelings." They prefer books on politics, history, and
science, where opinions can be supported by evidence. The more
polarized the opinions, the livelier the conversation—until frustration sets in. Then the most commonly heard arguments are
"Why can't you, just once, be reasonable" and "If only you would
be objective." The unstated subtext that drives these discussions:
"There is an optimal line of reasoning and I can know what it is."

In private conversations, these men are quite willing to acknowledge that a poet inherently sees the world differently than
an engineer, even that their own wives prefer novels to nonfiction.
And yet they persist in the belief that everyone should draw the
same conclusion if given the same information, as though reason
operated according to an obligatory physics, like the optics of an

eye. These book club members aren't alone. We are raised believing that reasonable discourse can establish the superiority of one line of thought over another. The underlying presumption is that each of us has an innate faculty of reason that can overcome our perceptual differences and see a problem from the "optimal perspective." One goal of this book is to dispel this misconception.

The process of reasoning arises from fundamental biological principles that we all share. But this is like saying that all computer programs arise from principles common to all algorithms. Even we computer illiterates know that Windows and Mac programs have the same generic structure—a series of algorithms—but that the programs are incompatible without additional bridging software. Which brings us to the question of the relationship between our code (our genes) and the formation of our thoughts. If the Windows-Mac analogy holds, we might suspect we share general powers of reason, but that our individual lines of reasoning for any given problem will be as idiosyncratic as our underlying code. In this chapter, I'd like to look at how genes might affect the very texture of our thoughts.

Before beginning this discussion, be assured that I am not promoting genes as the only, or even the primary determinant of our choice of thoughts. Though we tend to assign behavior into arbitrary categories, practical distinctions between nature and nurture are rarely possible. Genes and environment influence each other in a complex irreducible dance of positive and negative feedback. Nevertheless, if we want to understand why lines of reasoning cannot all be identical, we must consider how individual genetic makeup might influence our choice of cars, spouses, or presidents.

LET ME BEGIN by making an extraordinary and seemingly ridiculous proposition: Genes can affect our degree of interest in religion and spirituality. At first glance, such a suggestion feels preposterous; we see the lifetime pursuit of religion as a deliberate and intentional choice. If there is any single area of human thought over which we believe that we have control, it is in our ability to decide whether or not there is a God, a perfect hereafter, fire and brimstone, or that we are insignificant specks in a meaningless universe governed by chance.

But there's a huge problem with this presumption. Interviews of identical twins raised apart reveal a very strong correlation in the twins' religious attitudes and inclinations. If one twin is preoccupied with religious thoughts, the likelihood is high that his identical twin raised apart will have a similar inclination, and vice versa. (I am referring here to the degree of interest in religion and/or spiritual matters, not the choice of any particular religion.) Thomas Bouchard, University of Minnesota psychologist and head investigator of the most extensive and thoroughly evaluated group of identical twins raised apart, has even gone so far as to state that there is no evidence that parenting plays a substantial role in religious attitudes.

> A large and consistent body of evidence supports the influence of genetic factors upon personality. The evidence taken as a whole is overwhelming. We are led to what must for some seem a rather remarkable conclusion. The degree of monozygotic (identical) twin resemblance does not appear to depend upon whether the twins are reared together or apart.

> Our findings do not imply that parenting is without lasting effects. The remarkable similarity in social attitudes of identical twins raised apart does not show that parents cannot influence those traits, but that simply this does not tend to happen in most

families. This is true for a wide variety of social attitudes including religious interests.[1]

What if Bouchard is correct? What if the degree of our interest or disinterest in religion isn't primarily the result of parental and cultural exposure or metaphysical ruminations, but rather arises out of the sequence of amino acids comprising our DNA? Not possible, you counter, we are not genetic robots. A person highly spiritual by temperament might choose to reject all organized religions and become a card-carrying scoffer. Or he may become a secular humanist. People can "find God," or lose their "sense of faith." But what remains unclear is whether or not someone highly inclined toward the metaphysical can detach himself from or entirely subdue these spiritual yearnings.

A PERSONAL DIGRESSION: In writing my novel *Cellmates,* I reviewed Bouchard's data. Though there have been criticisms of the methodology, the studies seem well designed and the conclusions appropriate. My gut feeling has remained that Bouchard's studies point the way to some fundamental but puzzling truth. The obvious question is if DNA can influence how we think about religion, could it also be playing a role in my own idiosyncratic worldview?

From my earliest recollections onward, my thoughts have been colored by an overwhelming existential bent. Their origins are not apparent. Both of my parents were hardworking, practical, and resolutely nonphilosophical. Questioning was off-limits, even a bit scandalous. (Although there was occasionally a mischievous twinkle in my mother's eye, as if I was supposed to read between the lines of her frowns and her discouragement of anything but the most pragmatic musings.)

While in high school, I ushered at the local theater, the Actor's Workshop. Purely by accident, I saw the first San Francisco production of *Waiting for Godot*. I left the theater stunned. The resonance was unnerving, as though Beckett had slipped inside my head and written what I hadn't yet thought. Yes, this is how the world is. The pleasure was profound and comforting, as though I'd discovered a kindred spirit.

After fifty years, my admiration persists. More than any other artist (or neuroscientist), Beckett has captured the wondrous and amusing frustration of observing the mind in action. His "you must go on, I can't go on, you must go on, I'll go on," underscores the paradoxical and philosophically irresolvable relationship between thought and biology.[2]

Was being exposed to Beckett as an impressionable teenager a crucial element in how I now see the world, or was I biologically predisposed to appreciate his way of thinking? Was this pure nature or nurture or a mix, and how might I know?

Just before my mother died at age ninety-seven, I asked her what she had learned from her long life. Always circumspect and noncommittal on such subjects, she answered tersely, "So what?" I asked her again. "You must have developed some philosophy of life after all these years." She shrugged and repeated, "So what?" I persisted and asked again. She looked at me and said, deadpan and enigmatic, "I just told you what I learned."

In the hospital, her actual penultimate words: "In the end, I am only an ordinary person. No one special. No one to be remembered. Nothing."

After she died, I went to her apartment to clean out her few remaining belongings. In the back of her closet was a single cardboard box. Tucked beneath old photos and tax returns was a term paper on William James that I'd written in college. The opening paragraph, underlined in black felt-tip pen by my mother for emphasis, posed the same question that prompted this book—how do we know what we know? I do not remember writing the paper or ever discussing it with my parents. I can't remember ever showing them my college papers, though they did remain stored in their basement long after I'd moved away.

Nevertheless, there it was. Not only had my mother chosen to save and underline the key paragraph in this one paper out of all my papers through the years, but in the right-hand margin, next to her underlining, in feeble script, was written a single word—*yes*.

I HAVE NO way of determining if my particular philosophical approach to life has any genetic component. But if the identical twin studies have even a grain of truth, then this book may have been, at least in part, prompted by certain modes or styles of thinking arising out of biological predispositions. But how can DNA make Beckett more enticing than St. Thomas Aquinas, Wittgenstein more simpatico than Plato? In a recent review of genetic determinants of behavior, NIH geneticist Dennis Drayna has offered a provocative analysis of why some genes might be more directly related to behavior than others:

> More generally, human behavior is an exceedingly complex phenomenon and cannot be viewed as the product of a set of genes. Nevertheless, our behaviors that are instinctive and crucial to survival and reproduction are likely to be subject to simple genetic control. Such behaviors might include those necessary to maintain homeostasis—such as eating, drinking, excreting, and thermal

regulation—and those associated with mating and the maternal care of infants.[3]

At the top of the list of homeostatic behaviors would be the flight-or-fight response to a charging lion. An immediate no-thought-necessary reflexive reaction is clearly more adaptive than remaining defenseless while the cortex ponders, deliberates, vacillates, and/or waffles. If behavior crucial to survival is likely to be subject to simple genetic control, an ideal place to look for this correlation between genes and behavior would be in the amygdala—the site of origin for the fear response.

It has long been known that mice are easily conditioned in fear-avoidance responses. The typical conditioning response is associating the sound of a ringing bell with an electric shock to a mouse foot pad. Once conditioned, the response is hard to undo. This lifelong persistence of a conditioned fear response following a single period of conditioning has prompted LeDoux's observation that fear-generated emotional responses are persistent and indelible.

Recently a group of neurobiologists have determined that adult mice normally have a high concentration of a protein—stathmin—in the amygdala, but not in other areas of brain. By genetic manipulation, they were able to create knockout mice that lacked the ability to make this protein. (The term *knockout* comes from the selective inactivation of a single gene—the gene is referred to as knocked out.) Unlike normal mice, these knockout mice are difficult to condition to the fear response. They are strikingly less timid and readily explore new and unfamiliar environments within the lab—unlike their easily intimidated stathmin-loaded brethren. (Note the similarity to those patients with damaged or malfunctioning amygdala.) LeDoux's structural studies showing that destruction of the amygdala made animals less fearful have now been confirmed at a biochemical level. What once required gross anatomic destruction of an area of the brain can now be accomplished through precise manipulation of a single gene.

The researchers speculate that stathmin facilitates formation of fear-based memories that trigger unconscious avoidance behavior. By blocking the gene, the animals have a strikingly reduced ability to lay down fearful memories.[4] LeDoux has described this study as a major breakthrough, and even has suggested that we might one day have amygdala-specific therapy for treating anxiety states.[5]

Such studies support geneticist Drayna's observation that a profoundly adaptive mechanism—the fear response—is affected by a single gene. But how far can we go with this analogy? One of the problems in thinking about genetics and behavior is the difference between innate tendencies and actual predictability of behavior. Knowing that a mouse has a missing gene allows us to see what biochemical changes are manifest in the brain, but doesn't allow us to unfailingly predict what behavior will emerge. A mouse might be more prone to explore new environments, but the manner and degree will vary from mouse to mouse. A fearless but lazy mouse might appear as timid as the most fearful of his cage mates. What does emerge from such studies is a conceptual bridge between genes, thought, and behavior.

Alice in Genetic Wonderland, or Through Hyperbole's Looking Glass

Let me present a thoroughly implausible but nevertheless tantalizing hypothetical. Imagine that this same gene for encoding stathmin has been isolated in humans. Make the further unwarranted assumption that the gene can be manipulated so that it is either completely expressed or not expressed at all, and that its effect is not mitigated by other genes. (I am eliminating all real-life biological mechanisms responsible for varying degrees of gene expression.) You, a behavioral scientist, wish to study the effect of this gene on behavior. Through the miracle of Internet dating, you find and match a man who has full expression of the fear-response gene and a woman who is totally lacking in this gene. They become a couple. Neither has any awareness of whether or not they have such a gene or even if such a gene exists. (They are experiment naïve.)

To see if the gene can create a measurable effect on behavior, you ask them to plan a cross-country plane trip. Your goal is to see if a gene that affects the fear response will be a factor in how much time before plane departure each will want to leave for the airport. Presumably the husband will want to leave earlier to allow for unexpected traffic, check-in delays, and so on. In a preliminary interview you confirm that the husband and wife have different memories of prior flights. The husband immediately describes several prior hair-raising experiences, including being stranded overnight in the Timbuktu airport. His wife has no such thoughts. (Without the gene for storing bad memories, she will be a perpetual blank slate of optimism.) To record the different responses, you install a video camera at the breakfast table and take continuous audio and visual footage. As expected, the wife suggests leaving the house at the last possible moment. But, to your surprise, the husband immediately agrees. A close examination of his face reveals no conflict; the amount of time he takes to make the decision is so brief that you don't suspect any underlying apprehension. You conclude that the presence of the fear-response gene didn't affect the husband's decision or any observable behavior.

What you can't know is how the gene affected his thoughts in undetectable ways. An additional piece of history is that the husband's prior two marriages ended in bitter divorces with both departing wives accusing him of being cowardly, consumed with anxiety, and filled with self-doubt. His self-esteem is lower than his remaining Enron shares. The decision when to leave for the airport kicks his genetic predisposition to be fearful into overdrive, but not in a single direction. He is faced with two competing sets of disaster probabilities—getting to the airport late and missing the flight versus upsetting his new bride by revealing his cowardly neurotic ways. Both risk-reward probabilities are inputted into his hidden layer where they silently duke it out. If the fear of rejection is greater than missing the plane, the husband will quickly agree with his wife. His relief at not being criticized or laughed at might even block out his awareness of any underlying anxiety over missing the flight.

Though the gene played a major role in his decision making, it wouldn't be detectable. The problem that cannot be addressed is that if a gene creates counterbalancing desires and needs, it may not be seen in any final decision. This link between genes, thoughts, and behavior allows us to better understand how genes might cause identical twins raised apart to share similar social attitudes without requiring us to fall into the trap of advocating genetic determinism. In the Bouchard studies, the twins expressed how they feel and what they are interested in and attracted to. Such attitudinal studies tell us what the twins *want* to do (under ideal circumstances), not what they *will* do. So many of the arguments over free will and determinism fail to make this simple distinction. Desire and action are not synonymous. If we were to find a complex of genes that dictated the relative degree of interest or disinterest in religious and spiritual matters, we might see these tendencies reflected in how we think and what we think about, but not necessarily in any specific observable action. If the gene created conflicting beliefs, we might not even see the effect on our thoughts. They would be factors within the hidden layer, but wouldn't be consciously experienced.

(I know an avowed atheist who privately confesses to having once been a Pentecostal Born Again. It doesn't take much imagination to see how his Born Again and atheistic thoughts might arise out of a similar genetic predisposition but result in diametrically opposite conclusions.)

Why I Can't Play Poker

The list of genes affecting behavior is rapidly growing. One of the most personally intriguing is the gene associated with risk-taking and novelty-seeking, including a propensity for gambling.[6] (The gene produces a reduced reward system sensitivity to dopamine, but is referred to as a gene that promotes risk-taking. Presumably these higher levels of risk-taking are sought out in order to generate desirable levels of dopamine-derived pleasure.) A genetic contribution to the desire to gamble isn't surprising; at a gut level, we already suspect innate differences between those friends who will bet on anything, and those who can't understand why someone would sit for hours on a hard stool in a smoky room just to see three unappetizing cherries line up in a row.

The question is, if a single gene might prompt us to bet the farm on an inside straight, what might be its effects on the very formation of our thoughts?

As a lifelong poker player, I have spent considerable time

developing a winning strategy, yet I am not a great player. I have long suspected a variety of flaws, but haven't figured out a clear solution. With the recent popularity of televised poker tournaments where the viewers can see the players' hole cards at the start of each hand, the problem has become transparent. The players with the best overall results are those who aggressively make selective large bluffs, a style with which I have never been entirely comfortable.

People can talk of intuition, reading the other player, and all the intangibles that make poker so fascinating, but this cannot explain how many have honed their skills online, where there is no opportunity to read body language and tells. (Chris Moneymaker, the 2003 World Series of Poker champion, had never played in a live tournament before, nor had he ever been to Las Vegas.) Many of today's top players are quite knowledgeable in game theory and use computer simulations to develop complex calculations as to the best strategy for any given circumstance. For example, if, over an extended period of time, the amount won on all your bluffs in a specific situation will exceed what you lose when the other players call, you should always make the play (until ongoing calculations reveal that other players are catching on).

Here's the problem. Based upon both personal observations and computer simulations, I have concluded that this selective large bluff strategy is superior to always folding a bad hand. Unfortunately, although the strategy will tell you the approximate chances of other players calling, it cannot tell you precisely when. To make that calculation, I would need to see the other players' hole cards. Trying to figure out what the other players have turns out to be of less value than just making the large bluff periodically.

Easier said than done. When I recognize the optimal situation for such a large-sized bluff, my mind balks by first asking, "But what if the other player calls?" I do not will this thought into consciousness. I would prefer not choosing this question as a starting point for my consideration of what play to make. It simply appears in much the same way as I might jump back at the first sight of a coiled black garden hose. But there's plenty of time to reconsider. If it is a major pot, I can ask the dealer for additional time. Also, I can plan ahead for such circumstances—practice endlessly at home, give myself a pregame pep talk, and make cryptic notes on cocktail napkins that I can glance at during the game. I can even tell myself to ignore the initial negative thought, and be ready to combat it with my practiced decision.

Yet when the time comes, I am not able to pull the trigger. I tell myself that the strategy works *generally*, but might not work with *this* hand. I cannot, by thought, generate the *feeling of conviction* that the laws of probabilities are actually in effect and that betting a bad hand occasionally is preferable to always folding. I cannot convince myself that what I know to be correct is actually correct.

Most top players that I know have a different response. They tend to first think, "If I make a big enough bet, my opponent will fold." They are equally aware of the possibility that an opponent might call, but are comfortable with the larger picture that the large bluff bet is a winning strategy. One World Series of Poker champion scolded me for being timid, saying that the difference between us was that he was not afraid to go broke. Watch one of these high-stakes TV poker games, wait until someone makes a huge bluff, and check out your own feelings. If we know advance that this play is sound strategy, we shouldn't be surprised. But we are. We laugh, watch in awe and admiration, and wonder, "How do they do it?" A major factor in the immense popularity of these poker programs is the thrill of watching others make decisions that we know are correct, but cannot make ourselves.

Naturally, I fault myself for lacking courage under fire. I am fully prepared to accept an inherent cowardice as affecting my thinking. But there's a compounding problem. If I am lacking in the risk-seeking gene, instead of getting pleasure out of the large bluff, the very thought might trigger a mean case of nausea and the shakes.

The actual feeling of reward is more than just pure pain or pleasure and approach and avoidance. We might bluff to have a feeling of power, the joy of scooping in a big pot and stacking up our chips, or to experience the pure euphoria of a particular sequence of cards (the straight flush, for example). To provide this range of pleasures, the mesolimbic dopamine reward system is intimately linked to our entire emotional palette, including all of our feelings and mood states. At the top of this list and a necessary prerequisite is the *feeling of knowing*. First we learn the strategies, and then we can experience the joy of implementation. Ironically, it is this feeling state that others look for in your eyes when you make the bluff. Your projected *sense of conviction* helps convince the others that you aren't bluffing. The great poker players thrive on lesser opponents' lack of conviction—a neurophysiological dilemma for those who want to adopt new strategies without being fully convinced at a biological level.

TO DATE, THE studies of a gene's effect on the desire to gamble have focused on situations commonly perceived as "gambling." Volunteers are asked to play various games or make financial decisions based upon perceived risks; fMRIs record which areas light up and by how much. But what if the same gene affects questions not normally considered related to gambling? As an example, let's consider whether or not to open the Alaskan oil fields for unlimited drilling exploration. As soon as I pose this question, I am confronted with a clear risk-reward calculation: If we do open up the fields, can we subsequently correct any ecological catastrophes caused by the drilling? Before I can gather my thoughts, I see images of the *Exxon Valdez* oil spill and its effect on trapped wildlife. For someone else, the immediate reaction might be the image of long lines of cars waiting at the gas pumps during the major gasoline shortage in the 1970s. Neither of us consciously chose these initial compelling images that will shape our conscious decisions. The hidden layer has voted on what is most important and sent that image up into awareness—a calculation dependent upon all hidden layer factors, including genetic predispositions.

Think of the difference between two poker players—one without the risk-taking gene, the other with it. Both have the same information, but the one without the gene worries what will happen if his bluff is called while the other feels confident that he won't be called. Now make these two players politicians voting on oil drilling in Alaska. One will worry about all conceivable catastrophes, while the other will shrug off the downside risks with the added optimism that the miracles of modern technology can clean up any spills.

Or make them oncologists. A very good friend of mine developed non-Hodgkin's lymphoma. Standard chemotherapy failed; he went to the two local university medical centers to inquire about bone marrow transplant. Both oncologists indicated that the percentage increase in survival from the transplant was the same as the increased death rate from the treatment. Risk exactly equaled benefit. My friend was stumped and asked both doctors what they would do if they were the patient. Freed from statistics, and now only voicing personal preferences, each was quite convinced that he could make a recommendation. One voted yes; the other voted no.

From politics to medicine, seemingly deliberate reasons for a decision will be influenced by innate risk tolerances. A close look at most contentious issues of the day reveals the same problem. Arguments about capital punishment, abortion, stem cell research, cloning, and genetic engineering often are the reflection of differing risk-reward calculations. In thinking about capital punishment, a major consideration is one's degree of concern that an innocent man will be executed. For some, the slightest risk isn't acceptable; for others, it is. The arguments about genetic engineering often pit the slippery slope argument—"once we start down that road, there is no turning back" and "it will be the opening of Pandora's box"— against the acceptance of some degree of risk and the belief that "we have adequate controls and can fix any mistakes in judgment."

With such examples, it would be utter folly to attribute a decision solely to the presence or absence of a risk-seeking gene. On the other hand it would be equally shortsighted not to consider that genes do play a role. But as soon as we postulate genes and risk-taking, we immediately sense that the problem is more complex. Let's go back to poker. I might lack the risk gene, and be doubly cursed with maximum expression of the stathmin protein in my amygdala (I am risk gene negative and stathmin gene positive). Not only will I get less pleasure out of bluffing, but each time I contemplate this decision, I will instantly remember every agonizing loss when another player called my bluff.

Combining genes quickly produces exponential possibilities. To continue with our hypothetical oversimplification: What if future genetic testing were to demonstrate that the most strident advocates of environmental laissez-faire policies fell on the far end of the daredevil genetic spectrum? Such risk-taking gene positive, stathmin gene negative politicians could not be easily intimidated or humiliated; they wouldn't readily recall embarrassing or compromising situations. We bemoan our least favorite public official's total lack of self-awareness, but what if this apparent insensitivity is, in part, a function of a turned-off amygdala? No prior bad experiences would be triggered. Criticism wouldn't land anywhere. The politician could feel entirely justified in saying that he didn't understand what the fuss about global warming is all about.

I have been unable to fully assimilate the superior poker strategy of the random large bluff. I am aware that my choice of less profitable strategies isn't the optimal decision, but nevertheless is the one with which I am most comfortable. I am open to the idea that this flawed decision-making ability might even have a genetic component. Similarly, when thinking about environmental issues, I am aware that I see greater risks than the pro-drilling advocates. The unanswered and perhaps unanswerable question: If a conservationist has more stathmin, less risk-taking gene, and the pro-drilling advocate has less stathmin, more risk-taking gene, how can the two have a reasonable dialogue? Their basic genetic predispositions will create different lines of reasoning *and* an uneven playing field. The conservationist will more readily respond to innate fears and might be more easily intimidated.

BACK TO DR. Drayna's observation that genes most crucial for survival are the ones most likely to have a direct effect on behavior. Given the obvious survival benefits of the general category of the *feelings of knowing*, it wouldn't be surprising if such feelings also strongly correlated with genetic predispositions. Unfortunately, given the absence of a suitable animal model and the enormously complicated phenomenology of the *feeling of knowing*, it is unlikely that we will ever adequately sort out the genetic component. But we do have the soft data such as the identical twin raised apart studies showing familial clustering of attitudes toward religion and spirituality as well as other personality traits. The degree to which one seems inclined toward a state of certainty or doubt might itself be in part an expression of the ease with which one experiences a deep *feeling of knowing*. One day, the know-it-all and the perennial skeptic might be seen as the two extreme on-and-off positions of the gene(s) for the *feeling of knowing*.

BUT THERE'S A further complexity that we need to address; it is impossible to discuss genetic influence upon our thoughts without considering the wide-ranging effects of environment on gene expression. Genes do not operate in a vacuum. To put this into perspective, I'd like to briefly present a landmark study on effects of environmental sound on basic language acquisition and speech development.

The auditory cortex—the part of the brain that processes incoming sounds—is functionally arranged with specific regions being preferentially sensitive to a relatively narrow bandwidth of sound. By inserting microelectrodes into the auditory cortex of anesthetized rats, researchers can create detailed topographic maps of which areas process which frequencies. For any given frequency presented to the rat, one discrete area will fire madly while the remaining auditory cortex remains relatively silent.

This arrangement of the auditory cortex is like an elaborate card trick at a football game. The cheering section is subdivided into many microsections, each with its own set of flash cards and instructions. Individuals can only hold up a single card; by itself, this card is devoid of any specific message. If everyone does his job correctly, including reading the instruction sheet precisely, a meaningful pattern will emerge from the collective showing of all the cards. The auditory cortex works the same way—genes are the instruction sheet for each microsection.

Rats make a convenient model for studying brain development. The rat's auditory cortex continues to develop for about two weeks after birth. After that time, there is little further change. This initial window of brain plasticity allows researchers to study how environmental inputs might influence genetically programmed early brain development. If the cortex could be physically altered via environmental exposure, it would be a key insight into how the maturing brain is shaped by circumstance.

University of California, San Francisco, neuroscientist Michael Merzenich wanted to see if altering the environmental sound during the crucial period of postnatal brain development would change the anatomy of the auditory cortex. Merzenich designed an ingenious experiment in which he limited the sound exposure of a group of newborn rats to single frequency tones (monotones). After two weeks, when cortical development was largely complete, he studied the frequency response distribution within the rat auditory cortex. If genes were the sole determinant of brain development, the topographic map of the auditory cortex would be the same as in rats exposed to a normal range of environmental sounds. Instead, the neurons for the exposed frequencies were more plentiful and covered a far greater area of the auditory cortex than those neurons for frequencies the rats hadn't heard. The entire cortex had shifted to be maximally responsive to those environmental sounds present during the crucial stage of brain development.[7]

Merzenich's inference was that our brains are anatomically biased to preferentially hear what we are exposed to as young children. Conversely, we will have more difficulty hearing less frequently presented sounds during this crucial period of brain development. To test this hypothesis, he exposed another group of newborn rats to continuous moderately loud background noise (white noise). These rats demonstrated both delayed development of the auditory cortex as well as defective sound recognition.[8] The background noise interfered with optimal auditory development.

Assuming that these rat experiments are generally applicable to humans (the evidence is substantial),[9] consider the following hypothetical. Incoming English is a composite of approximately forty to forty-five phonemes. Via repetitive exposure and eventual pattern recognition, the brain builds neural networks that learn to detect individual phonemes and then a combination of phonemes. We begin with "da" and progress to daddy and Dada. From the very beginning of this trial-and-error process of language acquisition is the need for a suitable reward. Whether it's a pat on the head, a mother's smile, or "Yes, Virginia, that is a Z," the *feeling of knowing* becomes an integral and inseparable feature of the most basic neural networks for the recognition of letters, symbols, and phonemes. As a result, the most basic development of language will be influenced by the biases of those who are teaching us. What we are told is correct will shape all subsequent language-based thought. It is with these already colored building blocks of language that we listen to our teachers, choose our leaders, devise scientific experiments, theorize about philosophy and religion, and decide our futures.

Now consider what might happen if the incoming speech is garbled in the same manner that white noise alters the functional anatomy of a developing rat's auditory cortex. Picture a child trying to acquire language amid the background sounds of refrigerator motors, fans, air conditioners, hair dryers, a blaring TV, barking dogs, the close proximity of squabbling family members, the din of nearby street traffic, emergency sirens, and Alice Cooper coming from the apartment next door. Compound the problem by relegating the teaching of language to full-time working parents and/or harassed caretakers with limited time and vocabulary, improper syntax, and unusual pronunciations.

Is this a factor in America's declining literacy rates despite greater educational opportunities?[10] Are our children becoming the study group for the human equivalent of Merzenich's experiments, with cultural biases affecting structural brain development of the subsequent generation? As he has said, referring to inner-city children, "There's very powerful evidence that, at least in many children slow to learn language, the problem is a true delay in the development in the processing of their native language that leaves them with a defective language and their process is actually idealized not for English or Spanish, but for noisy English or noisy Spanish."[11]

What is most compelling about Merzenich's studies is the revelation of how the complex interaction of nature and nurture is present from the very beginning of brain development. Identical genetics will not result in identical brain structures. To put this into a more familiar context, let's return to the Windows-Mac analogy. Imagine seeing the exact same image on two monitors—one

Windows-driven and the other Mac-driven. Looking at this picture will tell us nothing about how these images were created. The successive lines of code—equivalent to lines of reasoning—will be different, yet the final image will be the same. As we saw with the couple preparing to leave for the airport, complete and full agreement is not synonymous with identical thought processes. Even when we fully agree on an idea, this agreement arises out of different ways of thinking, involving utterly unique genetics and personal experience. To expect that we can get others to think as we do is to believe that we can overcome innate differences that make each of our thought processes as unique as our fingerprints.

11

Sensational Thoughts

Thoughts are the shadows of our sensations—always darker, emptier, simpler than these.

—Friedrich Nietzsche

As we know, there are known knowns; there are things we know we know. We also know there are known unknowns; that is to say we know there are some things we do not know. But there are also unknown unknowns—the ones we don't know we don't know.

—Donald Rumsfeld

ONCE WE GET BEYOND THE SIMPLEST OF SELF-DEFINING SE-mantic thoughts, pure thought cannot resolve itself. The first time you escape a charging lion by running up a tree, reason will tell you that this is an excellent strategy. But eventually you learn from experience that great strategies sometimes fail miserably and that there might be better options that you haven't considered. The best that reason can do in the way of confirmation of the strategy is to declare that climbing the tree was effective *this* time.

As an isolated system, thought is doomed to the perpetual "yes, but," that arises out of not being able to know what you don't know. Without a circuit breaker, indecision and inaction would rule the day. What is needed is a mental switch that stops infinite ruminations and calms our fears of missing an unknown superior alternative. Such a switch can't be a thought or we would be back at the same problem. The simplest solution would be a sensation that feels like a thought but isn't subject to thought's perpetual self-questioning. The constellation of mental states that constitutes the *feeling of knowing* is a marvelous adaptation that solves a very real metaphysical dilemma of how to reach a conclusion.

In this chapter, I'd like to present a few examples of other seldom discussed mental sensations that are crucial to how we think about thinking. To make the point as provocative as possible, it is the goal of this chapter to show how thought cannot exist without sensation—both sensations from the outside world as perceived by the body and internal mental states.

TO BEGIN, CONSIDER how we think about the world in general. Imagine a very smart disembodied brain arriving by FedEx from some distant galaxy. Suspended in a jar, it is without sense organs—no eyes, ears, or peripheral sensations. The question is, how would it think about the world? This jar-brain could easily memorize the definitions for force, mass, and acceleration and the equation $f = ma$ without any personal experience of any of these conditions. But having never felt gravity's tug, it seems unimaginable that it could, from scratch, conceptualize the equation. We all know the probably apocryphal tale of Newton under the apple tree, watching the apple fall to the ground. Think of all of the prior experiences that went into understanding the simple observation that the apple weighed something, gained speed as it fell, and hit the ground with some calculable force. The isolated brain would never have experienced the bodily sensations that corresponded to the concepts of force, mass, and acceleration. It would never have floored a four-hundred-horsepower hot rod, felt the spinning tires, the sudden uncontrollable lurch forward, and felt its head snapping backward. It would not have any physical recollection of the surprise of picking up an seemingly light object—say a small ball—and being struck by its unexpected weight, before realizing that the ball was made of lead.

How would the jar-brain think about speed laws without some feeling as to what speed *is*? Or contemplate aesthetics. What would beauty mean to a disembodied mind? If you've never seen ugly, you cannot know what is beautiful. If you've never heard dissonance and cacophony, you cannot know when something is harmonious. We need a sensory appreciation of the world in order to give our thoughts palpable meaning.

In *Philosophy in the Flesh: The Embodied Mind and Its Challenge to Western Thought*, cognitive scientists George Lakoff and Mark Johnson offer a succinct summary.

> Reason is not disembodied, as the tradition has largely held, but arises from the nature of our brains, bodies, and bodily experiences. . . . The same neural and cognitive mechanisms that allow us to perceive and move around also create our conceptual systems and modes of reason. To understand reason, we must understand the details of our visual system, our motor system, and the general mechanisms of neural binding. *Reason is not a transcendent feature of the universe or of disembodied mind. Instead, it is shaped crucially by the peculiarities of our human bodies, by the remarkable details of the neural structure of our brains, and by the specifics of our everyday functioning in the world.*[1] (Italics mine.)

Disembodied thought is not a physiological option. Neither is a purely rational mind free from bodily and mental sensations and perceptions.

TO KNOW WHAT our minds are doing, we need some sensory system that can monitor our mental activities. Though my discussion has centered on the *feeling of knowing*, it is clear that there are also mental systems for monitoring self-perception. Perhaps the most universal, persistent, and unchallenged sensation is how your "self" feels as though it is located somewhere behind your eyes, somewhere inside your head, or at least somewhere within your body. It makes evolutionary sense that we don't normally feel ourselves "out there" in the cosmos or three blocks away in a saloon. Without a localized presence, you would be constantly "looking for yourself" without any guidelines for where "you" might be. If the sense of self is to have value in developing personal and social behavior, or even where to sit on the bus, we must know where "we" stand in relationship to others. Ideally, the brain would develop a global positioning system for the self.

Though no such single mechanism has been uncovered, recent research has shown that one area of the brain is instrumental in where we see our "self." The initial observation was made by a Swiss neurosurgical team performing direct cortical mapping on a young woman with uncontrolled epilepsy. During stimulation of the right temporo-parietal region, the patient consistently experienced "a sense of lightness, as if she were floating above herself. More remarkably, she seemed to see part of her body as if she were viewing it from the ceiling."[2] After eliciting the same out-of-body response in several other patients, the neurological team performed a simple follow-up experiment. They asked a group of volunteers to each imagine his "self" floating above his body. When they did, fMRIs showed definite activation of the same temporo-parietal region. But this fMRI response was limited to mental imagery of the self; envisioning other objects floating overhead activated different areas of the brain. The temporo-parietal region remained silent. Though subject to all the caveats inherent in correlating fMRI with behavior, the study has convinced the researchers that the temporo-parietal junction plays a major and specific role in how we sense where the self is located in relationship to the body.

It seems odd that we neurologists readily accept the idea of a peripheral proprioceptive system for determining the position of our body in space, yet only recently have we begun to postulate a similar system for localizing internal mental states such as the "self." The problem may lie in the very nature of how we experience our "self." Whether we see the self as a purely emergent brain function or an actual physical entity such as a material "soul," we sense that the self is a fixed point at the center of our consciousness and not a moving part in the way that a knee changes its position relative to the ankle. And yet, we must have some sensory system that tells us where "we" are located, or we wouldn't feel that we are present at all.

JUST AS WE sense where our mind "is," we must be informed as to what it is doing. Awareness that we are thinking is a sensation that happens to us; it is not a thought that we can consciously will.[3] We *feel* that we are thinking in the same way that we feel bodily activity. Those thoughts that don't reach awareness aren't felt as being actively thought. Which leads us to the larger question of the role of mental sensory systems in differentiating conscious from unconscious thoughts.

A recent personal example: I am lying in bed trying to remember the name of the comic strip that featured an alligator and a possum seated under a tree philosophizing. My wife can't remember either. Though I cannot consciously drum up the name, I'm reasonably confident that, if I "sleep on it," the answer will "occur to me" in the morning. I tell my wife that I'm going to run an experiment in which I will ask the same question of my unconscious that I have just been asking myself consciously, "What's the comic strip name?" After setting the wheels in motion for my unconscious to solve the problem, I drift off to sleep.

When I awaken, I am surprised to hear the word *Pogo* rising out of my early-morning reverie. Despite knowing that I "asked my unconscious" the same question that I asked myself consciously, I do not feel a sense of intention immediately preceding the arrival of *Pogo*. I do not feel that I "thought the thought." The answer feels distinctly different from conscious recall. The neurologist in me reminds me that there is no compelling evidence that these two modes of remembering are dissimilar, yet that is how I feel about them. A perfect example of cognitive dissonance—I am unable to viscerally accept what I know to be so.

This separation of the process of thinking from the awareness of thinking might seem unnecessary, even counterproductive, but for a moment consider the alternative. What if we experienced every thought process as it occurred? The chaos would be overwhelming. I heard only *Pogo;* I did not hear *Peanuts, Calvin and Hobbes,* or any other consideration that was rejected. At this very moment, you aren't aware of your own myriad unconscious ruminations. Imagine worrying about where your kids should go to college, when to get a haircut, and trying to remember a missing word in an old Pepsi-Cola jingle all simultaneously vying for your attention while you are trying to read this paragraph. In order to focus your full attention on immediate concerns, it makes sense to have nondirected, less pressing, or longer-range thoughts occurring in silence.

Most neuroscientists believe that conscious thoughts are the mere tip of a cognitive iceberg and that the vast majority of "thought" occurs outside of awareness.[4] If so, is the apparent difference between conscious and unconscious thoughts based upon differing physiologies or on how these thoughts *feel?* To get another perspective on the sensations of thought, I'd like to present a couple of brief thought experiments. As you read the two contrasting examples, ask yourself if your feeling about the examples is different than your understanding of them.

You are a university-based pharmacologist seeking a treatment for a very rare genetic disease. The standard approach is to seek out a specific responsible protein and model a theoretical drug to block the effects of this protein. On your personal computer, you input all the pertinent data—from the complete human genome to all prior research done on this and related diseases. You expect these extremely complex calculations will take considerable time. As you are busy with several other projects, and don't want to be bothered screening all possible answers, you program in a second set of instructions that allows the computer to predict the probability that a particular drug will be useful. Only those drug formulations that reach a certain predetermined likelihood of being effective will be shown on the monitor screen. Less likely answers will be automatically rejected.

Your personal computer is very slow, but you've kept it through the years because it is perfectly quiet. There's no annoying fan or hard drive noise; the LEDs have all burned out. When the monitor switches to standby, you can't even tell if it is on, let alone actively working on the problem. It is the perfect black box computer.

Bad news. Shortly after entering the question, your government grant is canceled because of "insufficient progress in a timely manner." Your lab is consolidated; your new projects are entered on networked computers down the hall. You stop using your trusty PC but keep it plugged in out of sight under your desk. Time passes. Eventually you forget about the project.

One morning, when you arrive at the lab, the long-darkened monitor is blinking. On the screen is the formula for a new drug. Underneath is the statement, "The theoretical likelihood of this drug being effective is 99.999 percent." You are excited by the result and bummed out that the grant was prematurely canceled. "They should have known that these massive calculations take time," you mutter to yourself. You consider rewriting the grant, confident that your previous programming did exactly what it was designed to do.

In this scenario, you don't feel that the computer has done anything out of the ordinary. It was merely following instructions. The absence of flashing LEDs doesn't suggest that the computer is processing information differently than when the lights worked. You don't sense having witnessed a miracle, feel a need to coin a new vocabulary, or invoke intuition to describe a computer running without notification that it is running. You aren't bothered by the length of time between posing a complex question and getting the answer. Nor are you surprised that the only answer that showed up on the monitor was one with a high likelihood of being correct. All of these conditions were anticipated.

Now change the scenario. You are a novelist contemplating writing a fat multigenerational novel with a huge cast of characters. You spend a few months consciously deliberating possible plotlines and narrative arcs, but the sheer number of permutations and combinations is overwhelming. Eventually you tire of the idea and move on to a minimalist tale that spans one day and two characters. You are relieved and forget all about the unwieldy larger project.

Years pass. Then, without any apparent instigating event, you awaken with the plot to your long-abandoned book appearing to you whole cloth. From introductory sentence to final denouement—it's all there in a rush of words and images. You are overwhelmed by the *sense of rightness* of the solution and by the absence of effort on your part. You tell your friends that you were possessed, and that the book was dictated to you by "higher powers." At book signings, in a slightly embarrassed tone, you talk of transcendence and intuition in your fiction writing. No matter how many times you revisit that moment, it remains incomprehensible, even "otherworldly."

A major problem in distinguishing conscious from unconscious thoughts is our inherent difficulty in assigning intention to thoughts occurring outside of consciousness. We all accept that motivation and intention represent complex interactions between what we consciously and unconsciously want. Yet when an idea appears without clear and immediate preceding effort, it doesn't *feel* intentional. With the *Pogo* example, I clearly asked my unconscious to solve the problem, but because of the elapsed time between question and answer, *Pogo* felt as though it effortlessly appeared "out of the blue." So did the plot structure for your abandoned novel.

Contrast this difficulty in feeling that unconscious thoughts are willful with our unquestioning acceptance that the silent computer is performing according to clear and specific intention even when we aren't aware of it in action (the LEDs aren't flashing). The difference is a function of our biology. We don't need to *feel* a computer's intention because we know what intention we programmed in and we accept that any delay in getting an answer is a function of processor speed and the complexity of the question. But with our thoughts, any significant delay between question and answer tends to strip the thought of a sense of being intentional.

How the brain creates a sense of cause-and-effect isn't known, but the temporal relationship must be crucial. We must experience cause as preceding effect. The closer this proximity, the greater the feeling of intentionality. If I stub my toe and get an immediate pain, I am pretty sure that stubbing my toe caused the pain. But if I stub my toe and get toe pain three weeks later, I am less sure of the cause-and-effect relationship. The more time that elapses, the greater chance there are other possible explanations. If I ask myself a question and get an immediate answer, the answer feels like an intentional response to the question. But the longer the delay, the weaker the feeling of intentionality is; "Yes, that's what I thought" gradually shifts to "It just popped into my head."

In the baseball chapter, we saw that the brain reorders the batter's appreciation of time in order to present a seamless view of the present. Basic physics of what the batter sees are overridden by neural mechanisms necessary for a coherent sense of cause-and-effect—the batter must feel that he sees the ball approach the plate *before* he begins his swing. Feelings of intention run into a similar problem. A sense of having pondered a question must be present in consciousness in close proximity to an answer in order for us to feel a clear cause-and-effect relationship. But we are intending to do a wide variety of things at any given instant. We are planning tonight's dinner, next week's lecture, a trip to the mountains, when to pay our taxes, get our shoes resoled, and when to turn on the TiVo. Having myriad dissimilar intentions simultaneously present in consciousness would create a chaotic and confused mind; attention would be scattered among all the questions being entertained. Not having all intentions simultaneously front and center in awareness creates the illusion that some thoughts aren't intentional, but simply "occur to us." It would appear that evolution has chosen the uncluttered mind at the expense of stripping the feeling of intention from unconscious thoughts.

HOW THE UNCONSCIOUS decides what should be delivered into consciousness is a matter of fierce debate. We needn't know the exact mechanism to realize that the decision must include a probability calculation. Let's go back to the computer example. In order to avoid getting a report of all possible compounds considered as potential drug candidates, you have programmed in a probability equation that only sends high-likelihood drugs into awareness (onto the monitor screen). This is the same process that neural networks use for pattern recognition.

Imagine teaching your young daughter the alphabet. You and Big Bird take turns patiently repeating the letter A while pointing to it in various formats—on a cube, a blackboard, a coloring book, and so on. When your daughter looks at an A for the first time, she might see an H. After repeated trials and your reinforcement of correct responses, these alternative interpretations no longer rise into consciousness. In saying that your daughter has learned to recognize an A, you are also saying that her unconscious mind can accurately calculate the odds of the image being an A versus an H or a tent with an arrow through it. The projection of the correctness of A into consciousness and the simultaneous rejection of other possibilities is analogous to the computer equation that restricted displayed answers to those with a reasonable probability of being correct.

To sense how this calculation evolves into the *feeling of knowing*, take a look at the following figure and try to decide if it is an A or an H.

A

An inability to decide is equivalent to your brain seeing this figure as approximately a fifty-fifty proposition. (A and H are equally likely.) If you chose A or H, your pattern recognition system has calculated that one is more likely than the other. Add in further clues and the probabilities change dramatically.

TAE CAT

With THE, you feel confident that the symbol is an H. With CAT, you reverse the probabilities. With both words, you feel a high degree of likelihood of being correct. I would even venture that many of you feel certain of your interpretation. The calculation of probabilities has been transformed into a *feeling of knowing*.[5]

It would be foolish to suggest that the *feeling of knowing* is present in the unconscious—an unfelt feeling makes no sense. The likely explanation is that unconscious pattern recognition contains a calculation of probability of correctness, which is consciously experienced as a *feeling of knowing*. The closer the fit between previously learned patterns and the new incoming pattern, the greater the degree of the *feeling of correctness* will be. A perfect fit is likely to result in a very high degree of certainty. A puzzling pattern that doesn't match with prior experience won't be recognized—the resulting low probability calculation might be felt as strange, unfamiliar, wrong, "not right," or not felt at all.

AS WE HAVE neither the investigative tools nor sufficient circumstantial evidence to know how thoughts emerge from neurons—either consciously or unconsciously—we are free to speculate as to any possible mechanism. Cognitive scientist Steven Pinker has coined the colorful but inexact phrase "mentalese" to refer to the symbolic processes that make up unconscious thought while simultaneously expressing our deeper lack of understanding of these processes.[6] But it seems highly likely that the basic mechanisms are the same—neural networks processing information (from sights and sounds to the most abstract thoughts). To postulate fundamental differences between conscious and unconscious thoughts would mean that the basic biology of cognition changes as thoughts move in and out of consciousness. But that would be like saying that your Prius changes into a Ferrari when you drive it out of the garage.

When writing a novel, you can feel the difference between writing "anything that comes to mind" and willful plotting where you consciously reject certain possibilities. When actively thinking, the censoring editor is in the on position; during unconscious thought it is mercifully muted. But this is only a difference of inputted information; some possibilities are consciously rejected while others are encouraged. From a neural network schema, the basic process of hidden layer processing of inputs remains the same—only the inputs have been changed by the conscious editor. Rather than opting for the dubious premise that "unthought thoughts" represent a different "way of thinking," why not consider cognition as a single entity that is subdivided into various ways of being experienced?

These felt differences are substantial. Conscious thoughts have the embedded sensation of willful effort and intention; unconscious thoughts lack this sensation. Conscious thoughts feel as if they are being thought; unconscious thoughts don't. Unconscious thoughts that reach consciousness have been prescreened and assigned a higher likelihood of being worth pursuing than those ideas that do not reach consciousness. Unconscious thoughts with a sufficiently high calculated likelihood of correctness will be consciously experienced as *feeling right*.

Intuition and Gut Feelings Are Unconscious Thoughts Plus the Feeling of Knowing

Welcome to two of the most misunderstood terms in popular psychology—*intuition* and *gut feelings*. For starters, look at how many misconceptions are packed into the brief definitions found on Wikipedia.

INTUITION

1. A quick and ready *insight* seemingly independent of previous *experiences* or *empirical knowledge*.

2. Immediate apprehension or *cognition*, that is, knowledge or conviction without consideration, thought, or inference.

3. Understanding without apparent effort.

GUT FEELINGS

1. Feelings or ideas formed without any logical rationale.

2. A deep-down conviction that something is so without knowing why.

Without considering the physiological relationship between mental sensations and thought, we are forced to draw some peculiar conclusions. What exactly is immediate cognition without thought? Is this some yet-to-be discovered brain mechanism whereby a thought occurs without any underlying thought process? And what kind of thought would occur without prior experience, including prior bodily sensations? (The belief in a disembodied rational mind isn't easily discarded.) And understanding without apparent effort? Isn't that a thought stripped of the feeling of it having been intentionally thought? The most on-the-money observation—that a gut feeling is a deep down conviction that occurs without any underlying sense of knowing why—is nothing more than the description of the *feeling of knowing* unaccompanied by the awareness of a precipitating thought or a specific line of reasoning.

Deep down conviction *is* the *feeling of knowing*. By understanding the relationship between this sensation and unconscious thoughts, we won't feel the need to create new categories of cognition. As we've seen with "mystical experiences," the spontaneous appearance of the *feeling of knowing* is often described as a moment of profound understanding. The power of this *felt knowledge* cannot be underestimated, even when it exists independently of reason or any confirming evidence. The comparison to intuitions is unavoidable; an intuition also is the appearance of the *feeling of knowing* without the awareness of a triggering line of reasoning or a conscious evaluation of available evidence. In the next chapter, we shall examine popular ideas about intuition. Right now, I just want to emphasize how recognition and discussion of the sensations of a thought are integral to any theory of mind.

To summarize: Thoughts require sensory information. A disembodied mind cannot contemplate beauty or feel the differences between deep love, infatuation, and pure lust. To avoid confusion and chaos, our brains have sensory systems that selectively tell us when we are thinking a thought. These sensory systems also determine how we experience mental cause-and-effect and intentionality. And they are instrumental in imbuing our thoughts with a sense of their correctness or incorrectness. Without the embedded sensation of being on the right track, a thought wouldn't be worth the mind it's printed on. For me, the evidence is overwhelming.

We know the nature and quality of our thoughts via feelings, not reason. Feelings such as certainty, conviction, rightness and wrongness, clarity, and faith arise out of involuntary mental sensory systems that are integral and inseparable components of the thoughts that they qualify.

12

The Twin Pillars of Certainty: Reason and Objectivity

When Levin thought about what he was and what he lived for, he found no answer and fell into despair; but when he stopped asking himself about it, he seemed to know what he was and what he lived for, because he acted and lived firmly and definitely....

Reasoning led him into doubt and kept him from seeing what he should and should not do. Yet when he did not think, but lived, he constantly felt in his soul the presence of an infallible judge who decided which of two possible actions was better and which was worse; and whenever he did not act as he should, he felt it at once.

So he lived, not knowing and not seeing any possibility of knowing what he was and why he was living in the world, tormented by this ignorance to such a degree that he feared suicide, and at the same time firmly laying down his own particular, definite path in life.

—Leo Tolstoy, *Anna Karenina*

Abandoning the Idea of Rationality Is Unthinkable

Perhaps the most daunting challenge for cognitive scientists is to portray the mind in a way that is both emotionally satisfying and yet reflective of its inherent limits. The major stumbling block that must be addressed: There is no isolated circuitry within the brain that can engage itself in thought free from involuntary and undetectable influences. Without this ability, certainty is not a biologically justifiable state of mind. If this limitation were easy to accept, this book would be finished. But abandoning or even qualifying the idea of the self-examining mind flies in the face of every facet of contemporary thought.

Introspection and belief in personal change are predicated on the ability to stand back and recognize when we are off base or out to lunch. When trying to cut back on excessive rumination, unwanted fears, or obsessive handwashing, we need a fresh point of view, not another voice from the same tainted circuitry. In medical school, when talk therapy was more popular, my psychiatry professors routinely encouraged us to enlist the cooperation of the portion of the patient's mind not involved in a delusion or hallucination. I cringe when I recall asking an acutely psychotic patient if believing that the FBI had bugged his radio "made sense." This is the same line of reasoning that led John Nash's colleague to ask Nash how he could believe in such nonsense as becoming emperor of Antarctica. Exhortations to just be reasonable are based upon this underlying assumption.

Any concept of free will assumes that we possess a portion of mind that can rise above the biological processes that generated it. Scientific inquiry requires this same piece of mind to objectively weigh evidence. Without this belief, the *feeling of knowing* wouldn't feel like *knowing*. Every time it arose, we would ask the same question: How do we know that this sense of knowledge can be trusted? Talking about the impossibility of a rational mind generates this general category in the same way that an atheist needs the concept of God in order to refute it. In short, relinquishing the idea of pure reason goes against the grain of how we lead our lives. From the *feeling of knowing* to a sense of personal agency, the presumption of at least a sliver of a rational mind is the glue of daily discourse, scientific discovery, and self-awareness. At the same time, it is the source of mental rigidity, resistance to new ideas, and serves as the justification for fixed belief systems. (To avoid repetition, I will refer to this belief—that we can step back from our thoughts in order to judge them—as the *myth of the autonomous rational mind*.)

Science is continually providing new and sometimes startling observations that go against the grain of common sense; we can't ignore these revelations just because they don't fit in with our present view of ourselves. Having spent more than forty years in medicine, I have never seen a situation in which dishonesty—no matter how well intended—was a long-term solution. Sooner rather than later, we need to face the music. We cannot continue to tell ourselves that these contradictory or undesirable aspects of the mind either don't exist or can be overcome through brute effort.

In order to affect any meaningful change, we need to have a simple screening tool. Since beginning this book, I have increasingly found myself asking a single question of any idea—be it the latest scientific advances, a pop psychology book, or personal opinions (mine as well as those of others): Is the idea consistent with how the mind works? By applying this question to some of the most important ideas of the day, we can quickly sort out the reasonable from the unreasonable. Just as I don't want to go to a dentist whose dental tray is loaded with pliers and vials of ether, I don't want to waste my time on ideas based upon outdated notions of how the mind works. My goal in this section is to kindle a new way of thinking about a variety of difficult issues.

To begin, let's look at some views of the rational mind, starting with two of the most popular psychology books in recent years—Daniel Goleman's *Emotional Intelligence* and Malcolm Gladwell's *Blink*. Rather than present each argument in its entirety, I've tried to focus on those elements that most readily underscore the discrepancy between how the brain works and how we wish that it would work.

Popular Psychology and the Myth of the Rational Mind

You jump into a lake to save a drowning child *before* you are aware of having seen the child. A motorcyclist unexpectedly brakes in front of you. You lean on the horn, furious *before* you realize that the cyclist has slowed to avoid smashing into an elderly Labrador limping across the road. In both circumstances, your action preceded conscious perception. You didn't *see* the child until you had made the leap into the lake. You became angry *before* you saw why the motorcyclist braked.

Whether talking about swinging at an approaching baseball before fully seeing it, or jumping in a lake to save a drowning child before being fully aware of why, conscious perception takes longer than unconscious reaction times. Combine this observation with the role of the amygdala in unconscious fear responses, and you have the makings of the hugely popular theory of emotional intelligence. Here's a capsule summary from Daniel Goleman, a Harvard-trained psychologist, on the example of jumping in the lake.

The limbic brain proclaims an emergency, recruiting the rest of the brain to its urgent agenda. The hijacking occurs in an instant . . . before the neocortex, the thinking brain, has had a chance to glimpse fully what is happening, let alone decide if it is a good idea. This circuitry does much to explain the power of emotion to overwhelm rationality. . . . LeDoux's research explains how the amygdala can take control over what we do even as the thinking brain, the neocortex is still coming to a decision. . . . It is in moments such as these—when impulsive feeling overrides the rational—that the newly discovered role for the amygdala is pivotal. [His point:] Our emotions have a mind of their own, one which can hold views quite independently of our rational mind.[1]

Emotional intelligence is a different way of being smart. It includes knowing what your feelings are and using your feelings to make good decisions in life. It's being able to manage distressing moods well and control impulses.[2]

These two minds, the emotional and the rational, operate in tight harmony for the most part, intertwining their very different ways of knowing to guide us through the world. Ordinarily there is a balance between emotional and rational minds, with emotion feeding into and informing the operations of the rational mind, and the rational mind refining and sometimes vetoing the inputs of the emotions. Still, the emotional and rational minds are semi-independent faculties, each reflecting the operation of distinct, but interconnected circuitry in the brain. . . . When passions surge the balance tips; it is the emotional mind that captures the upper hand, swamping the rational mind.[3]

Goleman repeatedly emphasizes the rational mind and its ability to recognize and control the effects of potentially harmful feelings on decision making. At first glance, this makes perfect sense. Anyone with a tendency to be impulsive, impetuous, or headstrong understands that being calm and dispassionate allows for clearer thinking than being anxious or angry. But Goleman is assuming that we can know which feelings are inherently detrimental and when they are adversely affecting our thoughts.

When was the last time that you experienced the *feeling of knowing* and said to yourself, "Hold it, you are being negatively influenced by an impulsive and unjustified feeling"? When we falsely identify Izzy Nutz's house as being the same one we saw twenty years ago, we cannot possibly know that the *feeling of knowing* is erroneous. It is because we have the feeling that we mistakenly pick out the house. The feeling is part of the neural network formed twenty years ago when we originally identified Izzy's house. Only after the thought is formally tested and proven to be wrong (a stranger answers the door), can we know that the *feeling of knowing* was misleading.

Though useful in emphasizing that unrecognized foul moods and emotions can impact clarity of thought, the theory of emotional intelligence ultimately sidesteps the crucial question of how we determine whether our thoughts are free of perceptual illusions and unsuspected biases. And the repeated assertion of a rational mind sounds suspiciously like a disembodied mind capable of pure thought without inputs from bodily and mental sensations. Despite these drawbacks, the theory's primary message—we can improve our reasoning by knowing when it has gone awry—is immensely appealing. As I write these lines, I have briefly entertained the idea that I can suppress my negative feelings about the theory of emotional intelligence and give it a fair assessment. Hardly.

Reporter to Yogi Berra: *"Have you made up your mind yet?"*
Yogi: *"Not that I know of."*

In his 2002 book *Strangers to Ourselves*, Timothy Wilson, a professor of psychology at the University of Virginia, presents a superb overview of the reasons why the unconscious mind is inaccessible to self-analysis: "The bad news is that it is difficult to know ourselves because there is no direct access to the adaptive unconscious, no matter how hard we try. . . . Because our minds have evolved to operate largely outside of consciousness, it may not be possible to gain direct access to unconscious processing."[4] Wilson suggests that we are better off by combining introspection with observing how others react to us, and deducing the otherwise inaccessible nature of our minds from their responses. If others see us differently than we see ourselves, we need to incorporate this alternative view of ourselves into our personal narrative. He warns us that introspection without looking outward at how others see us can actually be counterproductive.

If he's correct, the impasse between the necessity for self-awareness and the limits of our self-assessment abilities can't be overcome through more brute thought. In agreeing with Wilson, we are left challenging the commonsense and folk psychology understanding of ourselves, including knowing the degree to which we are consciously responsible for our thoughts and actions. Indeed, Wilson opened his book with the salvo, "It usually seems that we consciously will our voluntary actions, but this is an illusion."[5] The important point for our discussion is that Wilson's advice to readers is consistent with his understanding of brain function and our input-hidden-layer-output model. In essence he is arguing that we cannot see the hidden layer in action and that any attempt at self-awareness must accept this limitation.[6]

Observations by cognitive scientists like Wilson have thrown modern psychology into an existential crisis. What are we to make of our minds when the vast majority of cognition goes on outside of consciousness? Self-unknowability is akin to Wittgenstein's famous aphorism: "Whereof one cannot speak, thereof one must be silent." But emphasizing the limits of introspection and self-awareness isn't exactly an easy sell. What to do?

One of Wilson's biggest fans is *The New Yorker* staff writer Malcolm Gladwell. On his Web site and in his book's endnotes, Gladwell has praised Wilson's *Strangers to Ourselves* as "probably the most influential book I've ever read," and it was instrumental in his decision to write *Blink*.[7] Yet Gladwell ends up by assuring us that we can train our unconscious to make better decisions, and that we have the capability to know when we've made the best decision. I've included a few brief quotes from Gladwell's Web site and his introduction to *Blink*. His deeply rooted desire to believe in a rational mind leads to some extraordinary conclusions.

> It's a book about rapid cognition, about the kind of thinking that happens in a blink of an eye. . . . You could also say that it's a book about intuition, except that I don't like that word. Intuition strikes me as a concept we use to describe emotional reactions, gut feelings—thoughts and impressions that don't seem entirely rational. But I think that what goes on in that first two seconds is perfectly rational. It's *thinking*—it's just thinking that moves a little faster and operates a little more mysteriously than the kind of deliberate, conscious decision-making that we usually associate with "thinking."[8]
>
> Decisions made very quickly can be every bit as good as decisions made cautiously and deliberately.
>
> When our powers of rapid cognition go awry, they go awry for a very specific and consistent set of reasons, and those reasons can

be identified and understood. It is possible to learn when to listen to that powerful onboard computer, and when to be wary of it.

Just as we can teach ourselves to think logically and deliberately, we can also teach ourselves to make better snap judgments.[9]

Wilson's conclusions fit nicely with the input-hidden-layer-output model, but Gladwell's arguments boil down to believing that we can look at output (a spontaneously occurring idea) and infer both inputs and the hidden layer. In order to further prop up self-knowledge, Gladwell has arbitrarily subdivided unconscious decisions into intuition and gut feelings that "don't seem entirely rational" and those unconscious split-second decisions that do. But what does rational mean if you are using your own perceptions—"seem entirely rational"—as the criteria for deciding on rationality? Nevertheless, by declaring a segment of the unconscious as being free from emotions and feelings, he is able to conjure up a new category of mental process—the split-second, perfectly rational unconscious decision. He offers an evolutionary explanation and a separate, though unspecified, mechanism of action. "The only way that human beings could ever have survived as a species for as long as we have is that we've developed *another kind of decision-making apparatus* that's *capable of making very quick judgments based on very little information*."[10] (Italics mine.) But as we've seen with LeDoux's example of unconscious decision making—the reflexive jumping back at the sight of a coiled black object—sometimes this split-second judgment is correct, the object is a snake, and sometimes it isn't, the object is a curled-up garden hose. Just because we develop split-second decision making to enhance survival doesn't guarantee that these decisions are always correct.

Earlier, I suggested that cognitive dissonance tends to be resolved in favor of feeling over reason. Internal bias and a misplaced *feeling of knowing* routinely overpower and outsmart the intellect. What Gladwell knows and cites as true—Wilson's well-supported argument—cannot compete with his desire to believe in a rational mind. The result is that Gladwell is forced to ignore or sidestep basic biology to the degree that he ends up refuting Wilson's premise that inspired his book.

An even more fanciful notion of the rational mind is presented by Roger Schank, the founder and director of Northwestern University's Institute for the Learning Sciences and former director of the Yale University Artificial Intelligence Project. Schank accepts lack of self-knowability and rationality in our personal decisions but believes that we retain rational judgment in considering the thoughts of others.

> I do not believe that people are capable of rational thought when it comes to making decisions in their own lives. People believe that they are behaving rationally and have thought things out, of course, but when major decisions are made—who to marry, where to live, what career to pursue, what college to attend—people's minds simply cannot cope with the complexity. When they try to rationally analyze potential options, their unconscious, emotional thoughts take over and make the choice for them. Decisions are made for us by our unconscious; the conscious is in charge of making up reasons for those decisions which sound rational.
>
> We can, on the other hand, think rationally about the choices that other people make. We can do this because *we do not know and are not trying to satisfy unconscious needs and childhood fantasies.*[11] (Italics mine.)

Goleman believes in a rational mind that can know when it is being fooled. Schank sees the ability to be rational limited to the assessment of others. Gladwell extends the idea of rationality to some unconscious thoughts, but not others. These three highly knowledgeable authors are living proof that the very concept of rationality is dependent upon personal perceptions and beliefs in how the mind works. No amount of contrary scientific evidence—even if cited as source material—can overcome their innate biases as to the nature of rationality. In a moment we shall see how this same cognitive dissonance affects our understanding of objectivity. But first, I'd like to address the issue of intuition, gut feelings, and Gladwell's "split-second decisions."

Claims of ways for harnessing and improving subconscious decisions are big business—from audiotapes and CDs teaching how to crack the "intuition code" to books offering practical guides to inner knowing. There are courses on intuitive learning, healing, investing, selling, and managing. We are encouraged to "trust your instincts," "go with your gut," or in the parlance of poker, "get a hunch and bet a bunch." Even yesterday's Chinese fortune cookie told me to "learn to trust my intuitions." Yet announcing that unconscious thoughts can provide valuable insights is nothing more than brilliant repackaging of the obvious. All thoughts—the trivial, the brilliant, the mundane, the profound, the catastrophic, and truly dangerous—percolate up from the unconscious (the hidden layer). The issue isn't whether or not unconscious thoughts can be of great value, but in sorting out those that are from those that aren't.

One of the classic arguments for the power of intuition is the story of how chemist Friedrich von Kekule's vision of a snake grabbing its own tail led to the discovery of the benzene ring. This vision, by itself, is neither accurate nor inaccurate. Kekule might have interpreted his vision as suggesting that he should learn a new backbreaking yoga position or that he should host an orgy. But Kekule, an astute chemist, came up with a testable hypothesis—the formula for the benzene ring. No one questions that creativity is dependent upon flights of pure fancy and previously unimaginable new associations, metaphors, and visions. But paying more attention to unconscious thoughts doesn't guarantee a higher degree of accuracy. It was Kekule's interpretation of the vision that led to a scientifically verifiable hypothesis. There isn't a lab test for a vision.

The distinction overlooked by Gladwell is that the "logic of discovery"—the unconscious hidden layer activity that generates "gut feelings" and "intuitions"—isn't the same as the "logic of justification"—the empiric methods that we have developed to test our ideas. All kinds of ideas—good and bad—bubble up unexpectedly. Some will feel like "truths." For example, we can read a poem or watch a funeral procession and feel that we had a profound insight into the human condition. There is a logic to this process in the sense that the hidden layer has made a series of calculations that have produced a feeling of knowledge about the world. But this isn't the same type of reasoning that allows us to determine if coffee ground enemas will cure cancer, or if the *Challenger* is free of design defects.

We have no mechanism for establishing the accuracy of a line of reasoning until it has produced a testable idea. At his death, the mathematician Ramanujan's notebook was filled with theorems that he was certain were correct. Some were subsequently proven correct; others turned out to be dead wrong. Ramanujan's lines of reasoning led to correct and incorrect answers; by looking at his original thoughts, he could not tell the difference. Only the resultant theorems were testable. To call such hunches "perfectly rational," is to misunderstand the nature of rationality.

A further problem is that if a gut feeling is an unconscious thought *plus* a strong *feeling of its correctness*, then this feeling influences how we assess this thought. Consider a recent study suggesting that complex decisions are best made by the unconscious. The study data are entirely consistent with the present-day understanding of unconscious cognition, but the author's conclusion illustrates the very problem the study was trying to address.

Dutch cognitive scientist Ap Dijksterhuis and colleagues asked eighty people to make decisions about simple and complex purchases, ranging from shampoos to furniture to cars. In one of the tests, half of the participants were asked to mull over the information they were given and then decide which products to buy. The other half were shown the information but then were interrupted and requested to solve a series of puzzles. At the end of the puzzle session, the participants were asked to make a choice of products to buy.

According to Dijksterhuis, "We found that when the choice was for something simple, such as purchasing oven mitts or shampoo, people made better decisions—ones that they remained happy with—if they consciously deliberated over the information. But once the decision was more complex such as for a house, too much thinking about it led people to make the wrong choice. Whereas, if their conscious mind was fully occupied on solving puzzles, their unconscious could freely consider all the information and they reached better decisions."[12]

The problem is that a better decision is equated with those that the participants "remained happy with." But personal satisfaction isn't necessarily reflective of the quality of a decision. We are often enthralled with what subsequently turn out to be horrible choices. Ask the designers of the Edsel or the captain of the *Titanic*. We can't possibly know if the decision to buy a house is right or wrong. If the participants are all happy with their choice of a fabulous beach house in Malibu, the decision seems right until a torrential rain loosens up the soil and the house slides into the Pacific Ocean.

The study participants' positive emotional responses to their decisions may reflect nothing more than the inability to shake off the initial *feeling of correctness* that accompanied the decision into awareness. Nevertheless, such studies are seductive because they allow us to continue believing in the accuracy of our unconscious. *Chicago Tribune*'s headline summary of the study—IF YOU REALLY THINK ABOUT IT, TRUST YOUR GUT FOR DECISIONS—elevates a gut feeling into a self-fulfilling prophecy.[13]

Most world-class poker players respect unconscious cognition; many spend considerable time honing their split-second decisions and gut feelings. The best no limit high-stakes players are often the best readers of their opponents' minds. But a seat by the ATM machine tells a different story. Poker players regularly experience low probability events—a succession of low likelihood cards that can turn "a perfect read of the opponent" and a "sure win" into a "horrible beat." Because this possibility looms over every decision, prudent players only expose a portion of their bankroll to any given bet. They are smart enough to put limits on their trust of any individual decision—conscious or unconscious.

This discussion isn't about whether or not unconscious cognition should play a role in our decision making; without unconscious cognition there wouldn't be any conscious decision making. The issue I have with gut feeling, intuition, and split-second decisions is in believing that we can know when to trust them without having any criteria for determining this trust. A feeling that a decision is right is not the same as providing evidence that it is right. Which brings us to the discussion of the relationship between the myth of the autonomous rational mind and our understanding of objectivity.

We see only what we know.

—Goethe

MY WIFE AND I are among a small group of neurologists and psychologists attending a University of California at Berkeley neuropsychology seminar. The lecturer announces that he is going to show us a thirty-second video of two basketball teams, one team dressed in white, the other in black, three players to a team. Our assignment is to count the number of times the men in black uniforms passed the ball back and forth.

There is plenty of time for an accurate count, yet I count ten and my wife counts eleven. Most of the audience counted eleven, so I am wondering if my wife has once again out-observed me when the lecturer stops, asks the group if anyone has seen anything unusual in the video.

No response.

"Anything at all?"

A sea of shaking heads.

"How many saw the gorilla?" the lecturer asks.

No one raises their hand.

"You're sure there was no gorilla?"

Most nod, though they are concerned. They know there wasn't a gorilla, but there must be a point to the video.

The lecturer reruns the tape. Toward the end of the tape, a person dressed in a black gorilla suit walks onto the court, stops in the center of the picture, thumps his chest for about nine seconds, and then walks off. The players continue passing the ball as if nothing unusual had happened.[14] The audience laughs with amusement and embarrassment at not having spotted the gorilla.

I have no doubt that the image was recorded by our retinas. The failure of perception took place between the retina and consciousness, suppressed by an alternative intent. (The research team termed this *inattentional blindness*.) When our attention was redirected to looking for a gorilla, we had no trouble seeing it, but we might well have missed something else.

This gorilla study underscores how any choice of evidence depends upon the mind-set of the observers. Each of us in the audience told our unconscious what to look for. To carry this out with maximal efficiency, an implicit second instruction was sent to the unconscious—to downplay or ignore irrelevant visual inputs. As we can't anticipate all inputs to be considered, this latter instruction is open-ended. The unconscious has free rein as to what should or should not be seen.

Few believe that individual perceptions represent an exact correspondence to the outer world. We know better than to believe that observations arise out of a neutral dispassionate mind. We accept that the unconscious is loaded with unrecognized agendas, motivations, and complex ill-defined innate predispositions. We shouldn't be surprised by the gorilla study, and yet, as though we cannot believe our eyes, we persist with the faded notion of objectivity.

In the early 1800s, there was an ongoing scientific dispute as to whether or not it was possible to undertake a scientific study without some prior bias. Charles Darwin responded in a 1861 letter to a friend: "About thirty years ago there was much talk that geologists ought only to observe and not theorize; and I well remember some one saying that at this rate a man might as well go into a gravel-pit and count the pebbles and describe the colors. How odd it is that anyone should not see that all observation must be for or against some view if it is to be of any service!"[15]

Darwin doesn't equivocate or hide behind the myth of the autonomous rational mind; his straightforward acceptance of how observations occur is consistent with our understanding of brain function. He doesn't suggest that we can rid our minds of such biases. He proceeds with a full knowledge of his limitations—an extraordinary achievement and a profound lesson to the rest of us.

Contrast Darwin's intellectual humility with this prominent cardiothoracic surgeon's late-night TV claim that he had reduced his cardiac surgery complications by running his hands over a patient's "preoperative aura" (an alleged no touching technique for healing). "I was as surprised as anyone at the positive results. And, let's be perfectly clear; I went into this project without any a priori assumptions." If the surgeon didn't have any a priori assumptions, why would he have undertaken the project? He didn't study the effect of eating lasagna or reading the *National Inquirer*. For me the claim of no a priori assumption is a red flag to the likelihood of bias.

This surgeon isn't alone in his belief. A quick look at *Merriam*

Webster's dictionary definitions of *objective* and *to know* reveal the same problem.

> **OBJECTIVE:** expressing or dealing with facts or conditions as perceived without distortion by personal feelings, prejudices, or interpretations.

> **TO KNOW:** to perceive directly; grasp in the mind with clarity or certainty; to regard as true beyond doubt.

The misrepresentation of perception doesn't require further comment. The less obvious error is equating clarity with certainty. Clarity is an involuntary mental sensation, not an objective determination. Combining the limits of perception with recognition that the sensation of clarity of mind isn't a conscious choice should be enough to lay the idea of pure objectivity to rest. But we are not about to abandon common language. The continuing belief that we can strip our ideas of biases runs deep and isn't limited to those with a marginal understanding of science.

In this contest between objectivity and biology, Stephen Jay Gould comes as close as is possible to a reasonable middle road: "Objectivity cannot be equated with mental blankness; rather, objectivity resides in recognizing your preferences and then subjecting them to especially harsh scrutiny."[16] Gould refutes the idea of a mental blank slate that can observe without prejudice, and warns us to look under every mental rock to see what biases might have been overlooked. But "recognizing your preferences" brings us back to the strange loop of the mind judging itself. Though he knew better, and warned us about bias, Gould could not discuss objectivity without tacitly accepting some degree of the autonomous rational mind.

Even when demonstrating the power of unconscious bias on decision making, the prevailing tendency is to downplay the result. In a study on fMRI and unconscious bias, Drew Westen, an Emory University psychologist, looked at how partisan subjects processed negative information about their candidate versus the opposing candidate (John Kerry versus George W. Bush). Westen had expected different areas of the frontal cortex—"the rational regions of our brain"—to light up when pondering negative information about a subject's preferred candidate. Instead, increased activity was maximal in several areas of the limbic system while the frontal cortex remained relatively silent. Westen concluded that for partisan subjects, political thinking is often predominantly emotional. This not surprising conclusion still left Westen with the problem of how we can alter such biased unconscious behavior. Westen's conclusion is similar to Gould's "harsh scrutiny": "It is possible to override these biases, but you have to engage in ruthless self reflection, to say, 'All right, I know what I want to believe, but I have to be honest.' "[17]

I share the same wish as Gould and Westen. I have put great value on introspection and have tried to be particularly mindful of my own biases, especially when making medical recommendations. And yet, what began as a personal journal based upon self-reflection has ended up as a book underscoring the limits of self-knowledge. I want Gould and Westen to be correct, but realize that the best we can hope for is the perfect oxymoron—partial objectivity.

A very difficult question facing interpreters of modern neurobiology is how to juggle the need for self-examination with the knowledge that an unspecified percentage of such assessments will be flawed, sometimes with serious consequences. No one seriously doubts Socrates' maxim: The unexamined life isn't worth living. Self-assessment and attempts at self-improvement are essential aspects of "the good life." Yes, we should engage in ruthless self-reflection and harsh scrutiny, but we should simultaneously acknowledge that such introspection will, at best, only result in a partial view of our minds at work. Complete objectivity is not an option.

An all-too-common example of how introspection cannot overcome the biology that shapes our thoughts is the unshakable sense of low esteem and all-pervasive guilt felt by a depressed patient with a bipolar disorder. The patient looks into every aspect of his life and is utterly convinced that he is completely worthless; everything wrong with his life is entirely his own fault. No amount of counseling from friends can convince him otherwise. Because he is certain that his self-understanding is correct, he refuses therapy and jumps off the Golden Gate Bridge. Meanwhile, another patient with the same symptoms makes it to a psychiatrist and is placed on antidepressants. When his mood lifts, he realizes that his interpretation of his low self-esteem was erroneous.

Our reluctance to face the problems of the rational mind stems in part from the feeling that the mind isn't of the same category as the body. We don't expect to jump twenty feet high or to swim underwater for a week; we can easily feel our physical limitations. But we don't feel the same limits on our thoughts. For example, you feel free to accept or reject this paragraph. Acknowledging all the subliminal factors that influence this decision doesn't override the more powerful feeling that you are in control of your thoughts. In essence, we are programmed to believe in bootstrap theories of improving our minds. Our mental limitations prevent us from accepting our mental limitations.

As an alternative to pure introspection, Timothy Wilson suggests that we should become "biographers of our own lives, distilling our behavior and feelings into a meaningful and effective narrative."[18] His point bears repeating. If as modern neuroscience strongly indicates, the self is an ongoing personal narrative constructed by the very mind that is examining itself, introspection is analogous to interpreting a complex work of fiction. To get a view of oneself that is relatively "in sync" with one's unconscious motives requires a combination of close, detailed analysis, looking at the work from a wide variety of angles (including the views of others), and a broad background knowledge from one's personal and cultural history to the latest genetics of behavior. However, the overriding requirement is that any self-assessment be seen within the light of its biological constraints.

FOR MANY YEARS I have wondered why some bright, well-trained doctors would perform unnecessary surgeries, recommend the unproven, and tout the dangerous. My first inclination was to make accusations of greed, indifference, arrogance, or ignorance. Only since writing this book have I begun to understand how much of apparent malfeasance arises out of this same faulty belief that we can know with certainty when something unproven is correct. A powerful contradiction at the heart of medical practice is that we learn from experience, but without adequate trials cannot know if our interpretation of the value of a particular treatment is correct. Very few of us have kept detailed records of every observation and how it ultimately panned out. Rarely have our personal observations been subjected to independent scientific scrutiny. We readily acknowledge how memories are selective. Nevertheless, most of us have a strong urge to believe our observations are correct and universally applicable. Years of training and centuries of tradition have taught us that these observations are the essence of what make us good or bad doctors. To doubt your experience is to question your abilities.

A compounding problem is that to the extent that a sense of pride arises out of feelings of uniqueness or originality, we are divided in our motivation. We want to be known for having original ideas, inspired hunches, and gut feelings that make a difference. Indeed, a "well-honed sixth sense" is considered a measure of the good clinician. But being a good doctor also requires sticking with the best medical evidence, even if it contradicts your personal experience. We need to distinguish between gut feeling and testable knowledge, between hunches and empirically tested evidence.[19]

To conclude this chapter, I'd like to briefly present a few scenarios that highlight how frequently failure to understand the limits of what we can know becomes the basis for the perpetuation of erroneous or misguided medical information. To avoid excessive favoritism, I've taken examples from both alternative and traditional (allopathic) medicine. To begin, let's pick a controversial subject for which nearly everyone has some preconceived opinions. The chances are good that you already have some personal experience with some form of alternative medical therapy from acupuncture and chiropractic treatments to herbal remedies and glucosamine for joint pain. You probably have some sense of whether or not these treatments are of value. As you read each example, feel how your mind picks and chooses what it wants to believe. Ask yourself whether you find yourself rejecting certain ideas because they go against what you already "know" to be correct. Try to approach the next section as you did the kite description at the beginning of the book.

Complementary and Alternative Medicine

A PBS *Frontline* interview with Andrew Weil, M.D., discussed osteopathic medicine:[20] "Let's take the example of osteopathic manipulation for recurrent ear infections in kids. I wrote up my experience with an old osteopath in Tucson, who was a master of a method called cranial therapy. He would take a kid, one treatment of this very noninvasive, inexpensive method and they would never get another ear infection. I saw this again and again. So based on my experience, I have recommended that kids with ear infections should go to osteopaths and get this method done.

"After twenty years of trying to get the research community interested in this, we finally set up some tests of doing this with kids with recurrent ear infections. *We were unable in those tests to prove that this had an effect. The problem is, I'm sure there's an effect there.* We couldn't capture it in the way we set up the experiment. Part of the problem is that osteopaths have very individual styles of doing this. Were the osteopaths that we used, were they doing it right? Was it the same kind of method as this old man that I saw? I don't know." (Italics mine.)

Weil continues: "I reported one case of a woman who had advanced lupus. . . . She fell in love and the disease disappeared. . . . Now a skeptic might say, well the disease would have done that anyway, or she didn't really have lupus, or there's no connection. Fine, let them say that. *I know that there's a connection there."* (Italics mine.)

The purpose of presenting these paragraphs isn't to point out methodological flaws such as the lack of standardization of study design (not knowing what the osteopaths actually did), or unjustified conclusions such as not stating the follow-up time necessary to conclude that these children "never" had another infection. What jumps out at the reader is the more basic problem of the *feeling of knowing* in shaping and clouding one's judgment. Weil feels that he can be sure of a beneficial effect even after his study has produced a negative result. We've all had this feeling—an inherent difficulty in accepting that a result is contrary to what we expected (and hoped for). This is the junction where science and belief part company. Weil could have said, "I have a very strong hunch that this treatment works, but couldn't prove it." If his reasoning remained sufficiently convincing, he could design a new study to test his hypothesis. But until he has positive supporting evidence, he is only justified in saying, "I believe," not "I am sure." By understanding that a physician is making a recommendation based upon an unsubstantiated gut feeling, not solid scientific evidence, a patient can draw his own conclusions as to the value of the opinion. Such a recommendation must also specify potential risks of accepting a gut feeling recommendation over proven therapies. Untreated middle ear infections can lead to chronic problems from bone infection (mastoiditis) to permanent hearing loss.

Instead, Weil sidesteps discussion of risks and concludes that a negative study represents a flawed study rather than disproving his hypothesis. This is the same cognitive dissonance that allowed our creationist geologist to understand the evidence for evolution, yet reject the evidence. The pattern repeats itself. Weil acknowledges that fluctuations in symptoms are commonly seen with systemic lupus, yet he "knows" that there was a connection between the patient's improvement and her falling in love. Anticipating possible criticisms, Weil argues for his personal objectivity and rationality.

"I think that my views are balanced. I'm seen as being reasonable, commonsensical, and balanced. I think I'm fairly even-handed in my criticisms of conventional medicine, alternative medicine. I don't have an axe to grind for or against any particular system." And yet, he provides the following account: "My interest in so-called alternative medicine goes way back before medical school. My love of plants is something I got from my mother. That led me to be a botany major. I remember when I was a teenager becoming very interested in hypnosis and that started me on the path of inquiry about mind-body interactions. I began reading about alternative therapies when I was in college and wrote a paper about them. So these interests long predate medical school. When I finished my internship, it was very clear to me that I did not want to practice that kind of medicine. It just seemed to me that first of all it caused too much direct harm. And secondly, in general it didn't really get at the root of disease processes and change them."

Contrast Weil's claim of no a priori bias—"no axe to grind"—with his not-so-subtle references to "that kind of medicine" and "caused too much harm," and the allegation that modern medicine doesn't seek out the basic causes of disease. The combination of implicit trust in gut feelings, the *feeling of knowing,* and the ability of introspection to ferret out personal bias have resulted in a recommendation for an unproven treatment at the risk of preventing prompt treatment with proven techniques.

Weil isn't alone in his approach to medicine. From the same *Frontline* interview, this time with a major university pharmacologist specializing in oncological research: "I believe in the necessity of research, but I know that personal experience is the 'proof of the pudding.' "[21]

Or this interview with Russell Targ, a physicist, pioneer in early laser research, and cofounder of the Stanford Research Institute's investigation into psychic abilities in the 1970s and 1980s. Targ was diagnosed with colon cancer in 1985. In 1992, CAT scan and ultrasound studies suggested a metastatic recurrence of the colon cancer. Targ was advised to undergo evaluations and possible chemotherapy. Instead, he called Jane Katra, a spiritual healer he had met at a parapsychology conference the previous summer. "Acting on her intuition, Katra felt compelled to tell Targ that he was not sick, and that he should not empower that concept by saying he was sick, or that he had cancer. 'All we actually know,' she said, 'is that there were spots on some film.' With Katra's ministrations and recommendations of major lifestyle changes, Katra acted on the theory of 'changing the host so the disease could no longer recognize him.' Targ improved. He never undertook the prescribed chemotherapy, and six weeks later CAT scans showed that the tumor had resolved into something entirely benign. He has been fine ever since."[22]

The medical criticism of this miraculous tale is straightforward: There was no tissue diagnosis of a recurrence to warrant the claim of subsequent tumor resolution. (Abdominal CAT scans commonly show benign abnormalities that mimic malignancies.) My concern is the belief that pure intuition can warrant advising a cancer patient against further medical evaluations. Worse, this "spiritual healer" can lug around Gladwell's *Blink* and cite chapter and verse, pointing to Gladwell's claims for a perfectly rational unconscious and our ability to know when it is misleading us.

Consider this quote from *Blink:* "But what would happen if we took our instincts seriously? . . . I believe . . . that the task of making sense of ourselves and our behavior requires that we acknowledge there can be as much value in the blink of an eye as in months of rational analysis."[23]

I doubt that Gladwell had this intention in mind, but as long as we continue to mass market the belief that instinct can be the equivalent of months of scientific study, we will have physicians recommending worthless treatments because the physician "knows in his heart" that the treatment works. We will have physicians who are "sure" that love can treat lupus. We will have the medically unsophisticated make life-and-death recommendations based upon hunches and dreams. With one blink we will be back in the Dark Ages.

Imagine how different each of these claims would have been if intuition and gut feeling were acknowledged to be unconscious (and unproven) thoughts associated with a strong *feeling of knowing* rather than bona fide forms of trustworthy knowledge.

The conflict between alternative medicine and traditional medicine would be relatively easy to resolve if we acknowledged that each represents a different form of knowledge. By definition, alternative medicine encompasses those treatments that have not yet been proven effective by traditional medical techniques. Claims are based upon personal observations, gut feelings, hunches, suspicions, and as yet untested hypotheses; all are forms of "felt knowledge." If you want to know whether ginkgo biloba prevents Alzheimer's disease, you can run a control study. If it is shown to be effective, ginkgo biloba should be adopted by the medical community—it would make the transition from alternative to mainstream medicine. If it isn't shown to be effective, you are entitled to maintain your belief that it might work. But you should acknowledge that you have retained a hunch that isn't presently supported by scientific evidence. If you recommend ginkgo biloba to a patient, you have the obligation to inform them that your recommendation is based upon an unconfirmed belief. Ditto for cranial manipulation for recurrent ear infections.

Good science is more than the mechanics of research and experimentation. Good science requires that scientists look inward—to contemplate the origin of their thoughts. The failures of science do not begin with flawed evidence or fumbled statistics; they begin with personal self-deception and an unjustified *sense of knowing*. Once you adopt the position that personal experience is the "proof of the pudding," reasoned discussion simply isn't possible. Good science requires distinguishing between "felt knowledge" and knowledge arising out of testable observations. "I am sure" is a mental sensation, not a testable conclusion. Put hunches, gut feelings, and intuitions into the suggestion box. Let empiric methods shake out the good from bad suggestions.

BEFORE CONTINUING, ASK yourself how you would classify your own approach to medical problems. Would you classify yourself as scrupulously "objective" and likely to rely exclusively upon published data? Or do you favor alternative treatments and suspect that traditional medicine has missed many opportunities because of its narrow-minded provincialism? Are you a worrier or easily reassured? Do you tend toward exaggeration or minimization of your complaints? Do you have a hypochondriacal streak or are you generally stoic? And so on. . . . The questions are endless and sometimes difficult to answer, but they are necessary in order to provide and receive optimal care. Let's take a simple, seemingly straightforward, and extremely common complaint—chronic back pain. As you read the following case history, imagine yourself as the patient—what would you want and what would you believe? Then, as the treating physician, how would you respond to each new piece of information?

Some years ago, I consulted on Mr. Z, an extremely successful, mid-forties Chicago businessman who complained of several years of relentless back pain. He had seen a dozen different specialists; his physical examinations, laboratory tests, and several MRI and CAT scans were normal. There was no history of any injury, predisposing condition, similar family history, or personal problems that might cause stress-tension complaints. The patient was adamant that something dreadfully wrong was being overlooked.

Before proceeding, try to ask a question that isn't based upon some prior assumption of what causes back pain and what might relieve it. Listen to the way that you pose a question, and how you would decide what constituted evidence that would answer the question. Do you think of chiropractic treatment, magnets, gravity-inversion boots, deep tissue massage, relaxation tapes, piriformis muscle injection, Feldenkrais techniques, prolotherapy, or Pilates exercises? Do you think of a friend who recovered after receiving a treatment pooh-poohed by his other doctors, a relative that had an undiagnosed cancer because doctors failed to do a bone scan, or a neighbor who suffers from fibromyalgia?

Given the normal lab tests, X-rays, MRI and CAT scans over a several-year period, would you be satisfied that the odds of missing a potentially treatable illness are quite low, or is your risk tolerance such that anything less than certainty means you should repeat all the tests? If you do repeat the MRI and CAT scans and they are normal, will you be satisfied, or will you request less well-studied or experimental tests? And if those studies revealed an abnormality not seen on the MRI or CAT scans, which would you believe? Would you be willing to undergo a surgical procedure based upon controversial or unproven studies? If so, why? Would you demand a control study as evidence of the value of surgery, or would you accept the doctor's assurance of pain relief based upon his *personal experience*?

Could you come to the conclusion that further testing was not going to yield a definitive answer? Or would you continue to have that nagging thought: "There must be a reason; if only medicine had the proper diagnostic tools"? If the doctor raised the possibility of simple tension, would you respond with relief or frustration and annoyance? Would an armful of journal articles emphasizing psychological issues as a major component of chronic back pain be convincing? If not, what would constitute an objective evidence of stress?

I have chosen the subject of chronic back pain because it is one of the most common reasons for seeing a doctor. Acute back pain is fairly straightforward—usually due to ordinary strains from working in the garden or picking up a garbage can or your granddaughter. But once the back pain becomes chronic, the accuracy of diagnosis drops off dramatically. X-rays and MRI scans show all kinds of abnormalities, but the correlation is poor. Perhaps the most devastating summary statement—a *New England Journal of Medicine* editorial—suggested that the cause of most chronic back pain cannot be accurately determined.[24]

Read that last sentence again. Is that really possible? With all the technology at our disposal, surely we have some ideas as to the cause of a chronically sore back. Does this statement *feel* right or wrong? Logical or impossible to accept? Contrary to common sense and personal experience? And herein lies the problem. Even when we understand that experts either don't know or wildly disagree as to the causes of a condition, we feel that we can know the most correct answer—as if there is one. Which is why the subject of back pain has generated so many theories, unproven treatments, and so much unnecessary surgery.

Now switch roles. Try being Mr. Z's treating doctor. You've done everything that is reasonable and have no clear explanation for his pain. What do you do next? Would you resort to unproven or experimental tests, prescribe yet another round of muscle relaxants or anti-inflammatory drugs? As a last-ditch effort, would you explore the possibility that Mr. Z could have unrecognized stress tension, maybe even a purely psychosomatic (somatoform) disorder? Would you admit defeat and tell the patient that you cannot help him? Or would you tell him that you have no clear idea what is causing his pain?

Before continuing, make some tentative diagnosis and treatment plan, then ask yourself what is your felt level of confidence that you are on the right track.

Now let me provide an additional piece of history. Quite by accident, shortly after seeing Mr. Z, I ran into his close friend, who spontaneously told me of Mr. Z's early childhood. Mr. Z's mother had been stricken with a severe case of polio when Mr. Z was a few months old. From that time on, she had been confined to an iron lung. Mr. Z's father had buried his sorrow in business ventures, including trips that kept him away from home most of the time. The friend opined that Mr. Z's lack of physical intimacy with his parents was a major factor in Mr. Z's lifelong competitive drive.

I found out that Z had been the top squash player in his local athletic club for nearly a decade. His back pain began shortly after he was beaten by a new member of the club. Rumors circulated; most of the club members who'd known Mr. Z for years speculated that Mr. Z could not bear to lose. More surprising was that since quitting squash, Mr. Z had taken up golf, playing several times a week until he became a par golfer. And, yes, golf often aggravates lower back pain, and is a common vocational hazard for the touring pros. But no, according to Mr. Z, golf didn't aggravate his pain.

Is this new information enough to draw a connection between psychosomatic pain and a lack of physical intimacy during a crucial period of Mr. Z's early childhood development? Or is this a cop-out, an authoritarian medical establishment presumption— the doctor finds out something about a patient that fits with what he already suspects, and uses this fact as evidence?

Which is the better evidence, the absence of any apparent psychological complaints by the patient or his family, coupled with your inability to detect any problems, or a serendipitously uncovered, highly charged, early developmental history? To qualify as evidence, would we need to have "objective" measurements for each of these positions? Remember, this is how medical questionnaires come into being—they are an attempt to statistically quantify the subjective so observations can rise to the level of evidence. But what kind of evidence is a detailed psychiatric history that cannot disclose what the patient cannot remember, or consciously decides to omit, or unconsciously has blocked out?

If you do favor a psychological component to the pain, is this sufficient reason to expose a patient to the painful revisiting of a major early childhood trauma? Do you believe early childhood trauma results in hardwiring that cannot be overcome with recognition, or do you feel that better self-understanding will expose the source of the pain and allow it to be "talked away"? Can you make any prediction as to the most likely outcome of this revelation?

Now the ethical issue: You have no method for determining the likelihood that you are correct. And you do run the risk that stirring up the embers of Mr. Z's past can lead to overt depression or other unforeseen negative emotional consequences.

I have presented the case of Mr. Z to underscore the difficulties of believing that there can be a strictly rational approach to a common medical problem. In this scenario, scientific method alone cannot provide a right answer. It cannot even provide a single best line of reasoning—every step of the decision-making process is subject to unconscious bias of both the patient and the doctor. Matching doctor to patient under such circumstances is like trying to superimpose the patterns of two Oriental rugs. For the patient and the doctor to approach the problem with similar lines of reasoning, they need the very fabric of their two lives to line up in the same direction.

And yet, all is not lost. Mr. Z can still get good medical care. A thoughtful, compassionate, and wise doctor is more likely to give Mr. Z good advice than a rushed, insensitive, or poorly trained physician. The purpose of this chapter is to expose the limits of any concept of rationality or objectivity, not to suggest that all answers are equal and everything is relative. Some opinions are more likely to be correct than others. The art of medicine, as imperfect as it is, remains a useful tool in the same way that introspection can provide partial insights but not complete answers. Part of the art of medicine is in recognizing the limits of the art of medicine.

To provide the best care possible, we should know when we are basing our decisions on science and when they are based upon unsubstantiated experience, hunches, and gut feelings. But, as we've seen, we aren't reliable assessors of such arbitrary distinctions. The alternative is a middle ground—an attempt to base our opinions on as thorough a scientific understanding as possible, while simultaneously reminding ourselves and our patients that our information will necessarily have been filtered through our own personal biases, affecting our selection of evidence and even which articles trigger a *sense of correctness*. Once we've made this admission, we have stepped off the pedestal of certainty and into the more realistic world of likelihoods and probabilities.

> *"The weather report says there's a 70 percent chance of rain today."*
> *"Yes, but is it going to rain?"*

With medicine, the results of major mistakes are fairly obvious. My wife's aunt died from a skin cancer that was overlooked by her physician. Despite her aunt's repeated requests for a biopsy, the doctor insisted that the lesion was benign and a biopsy wasn't indicated. If he'd been anything less than utterly certain, a biopsy would have been performed when the cancer was quite treatable. To conclude this chapter, I'd like to briefly explore the notion of the moral difference between certainty and the highly likely.

Imagine that you and your spouse have both worked hard for many years and are six months from retirement. You have both invested cautiously and have enough money saved to live comfortably but not extravagantly. You get a call from your stockbroker saying that he has a sure thing—a stock is going public in the morning and it is guaranteed to double within the year. The extra money would allow you to travel first class and get that summer cabin in the mountains. You ask him how certain the guarantee is. He says, "One hundred percent guaranteed by our firm and backed by Lloyd's of London. There's zero chance of anything going wrong." With this scenario, you decide to invest your entire savings account.

How much would you invest if the broker said, "It has a 99.999 percent likelihood of success. It's almost a sure thing, but there are no absolute guarantees." Suppose that you have the gene for risk-taking while your spouse is much more conservative and frets over losing a dollar at a bingo game. The downside of losing might not bother you—you enjoy your job and wouldn't mind continuing on. And, as you love to be in action, you would relish riding the stock's moment-to-moment gyrations. On the other hand, your wife has had it with her job and can't wait to start community college watercolor classes. The two of you are undecided as to how much to invest. You phone your broker back and ask him to advise you.

Any decision that is less than certain and involves the lives of others has an inescapable moral dimension that extends to both expected and unforeseen consequences. A 99.999 percent guarantee is not just $1/100,000$ less certain than a 100 percent guarantee. It is not "almost certain" or "nearly the same as certain." It is the difference between no possible adverse consequences and the possibility, albeit remote, of personal and financial ruin. The broker has the obligation to explain the difference; the couple also has the obligation to understand this difference. Whether or not they hate statistics, they need to understand the fundamental difference between certain and highly likely.

This moral obligation also extends to those opinions in which certainty is implicit, though not specifically stated. A prime example is the prediction. A University of California at Berkeley professor and MacArthur Award–winning scientist recently claimed that "There will be ten billion people on Earth by 2100—and all of them can live comfortably if advances in energy-saving technology continue."[25] The statement sounds innocuous; a research scientist is expressing his opinion about the likelihood of advances in energy-saving technology. But there's a huge difference between highly likely and without a doubt. If the professor's calculations are wrong, the consequences could be catastrophic. How different his claim would be if he'd said, "According to my calculations it is quite likely that by 2100 the earth will be able to comfortably accommodate ten billion people. But there is a slight chance that I am wrong and that my calculations could lead to serious mistakes in population planning."

With simple situations, such as the chances of hitting blackjack or a coin coming up heads or tails, we can calculate exact odds. There is no such calculation for the possibility of errors of complex thoughts. Harsh scrutiny and ruthless introspection will not improve this calculation any more than concentrating harder on the basketball video without specifically looking for a gorilla will increase the odds of seeing the gorilla. We cannot calculate the chances of unforeseen consequences.

Here's an example of not seeing the gorilla leading to a flat-out denial of possible catastrophic climatic changes. A Canadian geology professor is quoted as saying, "I cannot see a mechanism that would bring the amount of fresh water required to actually cause the hydrological cycle to collapse. An increased hydrological cycle because of climate change and global warming doesn't cut it as far as I'm concerned."[26] He concludes, "It is safe to say that global warming will not lead to the onset of a new ice age."[27]

"Cannot see a mechanism" is analogous to not seeing the gorilla. "Safe to conclude" is the moral equivalent of a 100 percent guarantee. So is "all of them can live comfortably if advances in energy-saving technology continue."

Recognizing the limits of the mind to assess itself should be sufficient for us to dispense with the faded notion of certainty, yet it doesn't mean that we have to throw up our hands in a pique of postmodern nihilism. We thrive on idealized goals that can't be met. In criticizing the limits of reason and objectivity, I do not wish to suggest that properly conducted scientific studies don't give us a pretty good idea of when something is likely to be correct. To me, *pretty good* is a linguistic statistic that falls somewhere in between *more likely than not* and *beyond a reasonable doubt*, yet avoids the pitfalls arising from the belief in complete objectivity.

13

Faith

Welcome to the F Word

ALL ARGUMENTS ABOUT REASON AND RATIONALITY EVEN-
tually get down to what we can know versus what we take on
faith. But any discussion of faith is intimately related to the issue
of how we determine life's purpose.[1] By now it should be apparent
that deeply felt purpose and meaning are exactly that—profound
mental sensations. Though the underlying brain mechanisms that
create these sensations aren't known, the biggest clue comes from
those who've undergone "mystical" moments. A common thread
of such descriptions is the sudden and unexpected appearance of
a "flood of pure meaning" or an inexplicable *feeling of knowing* of
what life is about *without* the awareness of any preceding or triggering thought. Whether or not it is appropriate to use the word
faith to describe a feeling of "now I know why I'm here," or "this
must be what it's all about," it is impossible to overlook the shared
qualities of the *feeling of knowing*, a *sense of faith*, and feelings of

purpose and *meaning*. All serve as both motivation and reward at the most basic level of thought. All correspond to James's idea of felt knowledge—mental sensations that feel like knowledge. (This visceral *sense of faith* is not to be confused with the cognitive pot-pourri of conscious but unsubstantiated ideas that become arti-cles of faith, such as beliefs in religion, alien abduction, blueberries as a prevention for Alzheimer's disease, and a six-thousand-year-old universe.)

A second line of evidence comes from descriptions of when the feeling isn't present. Though not necessarily aware of when we feel purpose and meaning, we are nearly always aware of the sick-ening feeling when we don't possess them. This isn't an intellec-tual misapprehension; it is a gut sense of disorientation and a loss of personal direction. Rarely are brute mental effort and self-help pep talks able to rekindle the missing feeling. For most of us, we simply wait patiently, knowing from past experience that the feel-ing will return in its own sweet time. A lost sense of purpose is like a lifetime traveling companion that has temporarily wan-dered off on her own. Because this separation of intellect and felt purpose is so crucial to unraveling the misconceptions at the heart of the science versus religion controversy, I'd like to offer Tolstoy's brief description of an attack of melancholy that overcame him at age fifty. Of particular interest is his conclusion as to the inability of science and reason to provide a personal sense of meaning.

Tolstoy and the Biology of Despair

I felt that something had broken within me on which my life had always rested, that I had nothing left to hold on to, and that morally my life had stopped. An invincible force compelled me to get rid of that existence. . . . It was a force like my old aspiration to live, only it impelled me in the opposite direction.

All this took place at a time when so far as my outer circumstances went, I ought to have been completely happy. I had a good wife who loved me and whom I loved; good children and a large property . . . I was respected by kinsfolk . . . and loaded with praise by strangers. Moreover, I was neither insane nor ill. On the contrary, I possessed a physical and mental strength, which I have rarely met in persons of my age.

And yet I could give no reasonable meaning to any actions of my life . . . I sought for an explanation in all the branches of knowledge acquired by men. . . . I sought like a man who is lost and seeks to save himself—and I found nothing. I became convinced, moreover, that all those before me who had sought for an answer in the sciences have also found nothing. And not only this, but that they have recognized that the very thing which was leading me to despair—the meaningless absurdity of life—is the only incontestable knowledge accessible to men.[2]

Today most psychiatrists would label Tolstoy's experience a depressive reaction; one of the hallmarks of severe clinical depression· is a diminished or absent sense of meaning and purpose. Most would suspect an underlying. neurotransmitter imbalance and prescribe selective serotonin uptake inhibitors (SSRIs) such as Prozac or Zoloft. Few would suggest a Norman Vincent Peale–style *God Helps He Who Helps Himself* audiotape or the "stiff upper lip" British approach. We don't browbeat depressed patients to "get over it" because we are willing to accept that brain chemistry aberrations somehow result in a loss of a sense of meaning. But when a sense of purpose and meaning is present, it isn't normally described as arising out of properly functioning neural mechanisms. Instead, purpose and meaning are discussed in metaphysical or religious terms. (I suspect that, if pressed, most of us would consider purpose and meaning to be conscious choices, or at least that they have a major volitional component.)

If we abandon the belief that the feelings of purpose and meaning are within our conscious control, and see them as involuntary mental sensations closely related to the *feeling of knowing*, we have a potentially powerful tool for reconsidering the science-religion conflict.

Caution: Deconstruction Zone Ahead

To represent the prototypic rationalist scientist stance, I've chosen its most persuasive and relentless spokesperson, Richard Dawkins, Oxford Professor of Public Understanding of Science. Two of his most famous quotes quickly illustrate the problem of believing that we can rationally choose whether or not to be religious.

"Faith is the great cop-out, the great excuse to evade the need to think and evaluate evidence. Faith is belief in spite of, even perhaps because of, the lack of evidence." And, "I will respect your views if you can justify them. But if you justify your views only by saying you have faith in them, I shall not respect them."[3]

When I read recommendations for cobra venom injections as the definitive treatment for multiple sclerosis or hear someone insist that a blastocyst has a soul, I feel compelled to ask, "Where's the evidence?" When terrorists fly planes into the World Trade Center, I am horrified by the power of religion to subvert the minds of the young. One of the overriding fears of our time is that excesses of belief may destroy civilization. So, at first glance,

Dawkins's criticism of faith-based arguments is right on. But can we follow his advice and still get up in the morning? Is it possible to have a sense of meaning and purpose without some feeling of faith?

Richard Dawkins candidly admits that he cannot live without some element of meaning. "They say to me, how can you bear to be alive if everything is so cold and empty and pointless? Well, at an academic level I think it is—but that doesn't mean you can live your life like that." His solution runs headlong into the problem he's devoted his career to debunking. He continues, "One answer is that I feel privileged to be allowed to understand why the world exists, and why I exist, and I want to share it with other people."[4]

Dawkins both believes in his powers of introspection and self-assessment and that he is mentally capable of understanding why the world and we exist—the myth of the autonomous rational mind. This is coupled with another act of faith—the belief that possessing complete knowledge of the physical laws of the universe will tell us why we are here. It is an extraordinary proposition to believe that an intellectual understanding of physical properties can reveal subjective metaphysical truths. Why we exist is a matter of personal opinion and speculation, not a question for scientific inquiry. An additional and even more basic problem is that Dawkins assumes that understanding why we are here is either synonymous with purpose or at least will trigger a sense of purpose and meaning. But reason isn't necessarily capable of summoning a sense of meaning—as Tolstoy so elegantly reminds us. Dawkins isn't even able to sidestep the language of religion that he's criticizing. Being *allowed* suggests the presence of a higher power that can grant this privilege. But who is granting this privilege if there is no higher power? Since Dawkins is a self-proclaimed atheist, I presume that he's referring to an all-powerful rational mind capable of this understanding. In essence, Dawkins is deifying the rational mind that will allow him to understand why he exists.

Dawkins conveniently illustrates the rationalist's dilemma: How do you articulate a personal sense of purpose when you intellectually have concluded that the world is pointless? What is the purpose of pointing out pointlessness? What does it mean to find purpose in understanding purposelessness? Once again we are back at the conflict between Dawkins's intellect (the world is pointless) and his mental sensation of purpose (I will show others that faith is irrational). To understand the intensity of this felt purpose, Google Dawkins's bio and speaking engagements. His near-evangelical effort to convince the faithful of the folly of their convictions has the same zealous ring as those missionaries who feel it is their duty to convert the heathens.

There is a problem basic to the science-religion controversy: Although the sense of purpose is a necessary and involuntary mental sensation, it isn't easily comprehensible solely as a sensation. It doesn't feel right to say, "I have a sense of purpose but don't know what it is." In order to think about purpose and meaning, we need labels. We attach words to spontaneously occurring feelings in order to incorporate them into a larger worldview. If we didn't use such language, the expression of purpose would be difficult if not impossible. If you doubt this, try to state your purpose or the meaning of life without expressing thanks, gratitude, obligation, moral imperative, or a need for a greater understanding of the unknown. Whatever the explanation there is an underlying implication of a something beyond us that needs to be acknowledged or pursued—from an all-knowing God to the awe-inspiring physical laws of the universe. Religious purpose might be described as a movement toward the understanding or embracing of a higher power. Scientific purpose might be described as a movement toward understanding the nature of the mystery of the universe.

How different the science-religion controversy would be if we acknowledged that a deeply felt sense of purpose is as necessary as hunger and thirst—all are universally necessary for survival and homeostasis. How we express these sensations will be a matter of personal taste and predilection. Some respond to a sense of thirst by wanting Gatorade while others opt for champagne. Neither choice is strictly "reasonable." A middle-of-the-night hankering for pickles and ice cream isn't a bizarre belief system—it is a hidden layer computation that includes the condition of pregnancy. Only by understanding that such seemingly peculiar tastes can be rooted in biology are we able to get up at three in the morning and trot down to the local 7-Eleven without thinking that the wife has gone off her rails.

Imagine the sense of purpose as a powerful committee member within the hidden layer. It carefully weighs all inputs, positively weighting those experiences and ideas that *feel right* while negatively weighting those that *feel wrong, strange,* or *unreal.* The best that a rational argument can accomplish is to add one more input into this cognitive stew. If it resonates deeply enough, change of opinion might occur. But this is a low probability uphill battle; the best of arguments is only one input pitted against a lifetime of acquired experience and biological tendencies operating outside of our conscious control. To expect well-reasoned arguments to easily alter personal expressions of purpose is to misunderstand the biology of belief. If there is to be any rapprochement between science and religion, both sides must accept this basic limitation.

Purpose reminds me of parents naming a newborn daughter. Prior to making the choice, the nameless child could be anyone—an essential aspect of her being hasn't yet been declared. After being named Alice, the baby is now identified as not being any other child and is distinct from all the names that weren't chosen. The baby is now Alice as opposed to someone else. Purpose begins as a nameless but essential mental state, but ends up being expressed via a variety of labels and justifications depending upon one's constitution and experience.

Dawkins's stated purpose is to discover how the world ticks. Stephen Hawking once said, "My goal is simple. It is complete understanding of the universe, why it is as it is and why it exists at all."[5] I presume that both men have a strong felt sense of purpose onto which they have grafted their belief in the rational mind and its unlimited capabilities. Others with different genetic predispositions, backgrounds, experience, and subjective self-assessments might interpret the same basic mental sensation as being evidence for the existence of God. Whether we opt for science or religion or both, we are telling ourselves stories about ourselves and the world in which we live. Stated purpose is a personal hidden layer-based narrative—not a reasoned argument.

TO SEE HOW a personal interpretation of a visceral sense of meaning affects the most seemingly neutral aspects of thought—pure numbers—let's look at one of the core assumptions of intelligent design theory.

Consider this quote from a 2003 article by Paul Davies, who is a Cambridge-trained physicist, a former professor of natural philosophy at the University of Adelaide in South Australia, and winner of the 1995 Templeton Award for advancement of the dialogue between science and religion.[6]

> It is hard to resist the impression that the present structure of the universe, apparently so sensitive to minor alterations in the numbers, has been rather carefully thought out. Such a conclusion can of course, only be subjective. In the end it boils down to a question of belief. . . . The seemingly miraculous concurrence of numerical values that nature has assigned to her fundamental constants must remain the most compelling evidence for an element of cosmic design.[7]

Davies concedes that his conclusions boil down to a question of belief, yet he calls his subjective interpretations of "seemingly miraculous concurrence of numerical values" compelling evidence. As a highly regarded theoretical physicist, Davies knows that the value attached to a number is a subjective interpretation, not evidence. The number 3.14 doesn't necessarily stand for pi; it can just as easily be the odds of making a flush, the latest version of your computer's screensaver, or the change left over after paying for a pizza. Nothing can be deduced simply by looking at the number—especially not meaning and purpose. If the odds of winning the lottery are one in a billion, winning the lottery doesn't tell us anything about why we might have won the lottery. The whole argument of luck, coincidence, miracle, or divine intervention hinges upon one's personal view of low-probability events. Yet, for Davies, the most convincing evidence for cosmic design arises out of his belief that low-probability events don't occur on their own.

The difference is that rationalists and skeptics see coincidence irrespective of the improbability of chance occurrences. Those inclined toward belief in higher powers see a finite point when coincidence becomes evidence for the miraculous. In a way, this isn't surprising. Most of us have our own personal relationships with numbers. If we buy a lottery ticket and lose, we don't think we are victims of a meaningless universe. We have no problem accepting that the odds are against winning. But if we do win, it is common to feel some sense of being "singled out" or "chosen." We might see ourselves differently than those who didn't win. If you come down with a cold that is sweeping the neighborhood, you don't think of yourself as a victim. But if you get a rare illness, it is hard not to ask, "Why me?" We have an innate tendency to characterize the unexpected and unlikely according to our worldview.

A similar problem plagues the interpretation of randomness—another major stumbling point between science and religion. The boiled-down argument is contained in Nobel laureate Steven Weinberg's famous quote from *The First Three Minutes:* "The more the universe appears comprehensible, the more it also appears pointless."[8] The underlying assumption is that the presence or absence of purpose can be determined based upon whether or not the universe evolved in a random manner. Randomness is an observation; it isn't evidence against a higher-order design. If I want my garden to look like a jungle, my best chance is to let the plants crawl all over one another. The garden may look like utter chaos, but that was my intent. Perhaps we are a well-designed experiment in futility.

The belief that we can rationally determine the difference between purpose and pointlessness arises out of a misunderstanding of the nature of purpose. We are further burdened by having a brain that learns by seeking generalizations over ambiguity. This preference prods us by producing its own mental state—the uncomfortable feeling that an ambiguous situation *must* have an answer. I suspect that this feeling is a prime mover in the science-religion debate. No matter how strong the evidence for our inability to know why we are here, we continue to search for an answer. Even when these questions arise out of paradoxes generated by contradictory brain functions, we *feel* that we should be able to solve the problem. The result is that we see patterns where none exist and don't see patterns that might exist. Combine our urge to categorize with an inherent tendency toward religiosity and it is not surprising that we will see a higher purpose rather than coincidence in low-likelihood events. Conversely, an innate skepticism and lack of spiritual tendencies is likely to favor the declaration that all is random and therefore pointless.

If these arguments were merely academic differences of opinion, they might be dismissed as irrelevant musings. But such arguments form the basis for major social decisions. Listen to Leon Kass, M.D., Ph.D., and the chairman of George W. Bush's Council on Bioethics.

We, on the other hand, with our dissection of cadavers, organ transplantation, cosmetic surgery, body shops, laboratory fertilization, surrogate wombs, gender-change surgery, "wanted" children, "rights over our bodies," sexual liberation, and other practices and beliefs that insist on our independence and autonomy, live more and more wholly for the here and now, subjugating everything we can to the exercise of our wills, with little respect for the nature and meaning of bodily life.[9]

Kass is convinced that his ability to know "the nature and meaning of bodily life" is so absolute as to exclude the possibility of valid alternative beliefs. Based upon this faith-based determination, Kass has been instrumental in the present administration's opposition to expand stem cell research.

THE SCIENCE-RELIGION CONTROVERSY cannot go away; it is rooted in biology. If we were to ban all discussions of religion, burn all religious books, even strip all words related to religion and faith from the dictionary, we would not eliminate religious feelings. Knowing that the sense of self is an emergent phenomenon arising out of simpler neuronal structures doesn't and won't stop theologians and philosophers from debating issues that they have no chance of resolving. Scorpions sting. We talk of religion, afterlife, soul, higher powers, muses, purpose, reason, objectivity, pointlessness, and randomness. We cannot help ourselves.

If, for most of us, science either is too complicated or cannot provide the heartfelt joy and meaning of religion, it is only natural that we will look elsewhere. Most scientists will privately admit that they are capable of understanding less and less of an increasingly more complex picture. Researchers in adjacent labs rarely understand one another's work; a good understanding of unallied fields is out of the question. For those less well versed in science, the gulf is even wider. It is certainly understandable that those of us who don't have a deep grasp of science might find awe in unfathomable mysteries, but won't embrace the notion of the eventual unraveling of the universe's mysteries as the reason for living. Even scientists aren't always "convinced." In a stunning about-face, Francis Collins, M.D., director of the National Human Genome Research Institute since 1993, went from being an

avowed atheist to an Evangelical Christian. When recently interviewed on PBS, Collins provided the following account of his
conversion:

> I was on a trip to the Northwest, and on a beautiful afternoon
> hiking in the Cascade Mountains, where the remarkable beauty
> of the creation around me was so overwhelming, I felt, "I cannot
> resist this another moment. This is something I have really longed
> for all my life without realizing it, and now I've got the chance to
> say yes." So I said yes. I was twenty-seven. I've never turned back.
> That was the most significant moment in my life.[10]

It is hard to imagine a more concise and moving description of
the struggle between his prior belief "that all of this stuff about
religion and faith was a carryover from an earlier, irrational time,
and now that science had begun to figure out how things really
work, we didn't need it anymore," and the recognition of his contrary, lifelong, deep-seated religious urges.

In a 2006 cover story for *Time* magazine, Dawkins and Collins
debated the presence of God. Despite the well-reasoned dialogue, neither changed his opinion. Given the powerful involuntary nature of the religious urge and the likelihood of underlying
genetic differences in our propensity toward religious feelings, we
shouldn't be surprised. Though I would prefer a world free from
(or at least unaffected by) fundamentalist beliefs, I can't see where
fundamentalists are going to abandon religion because scientists
portray a cold sterile world where all is pointless and faith is not
to be respected. Is this proclamation by Harvard Professor
Richard Lewontin likely to sway the religious to abandon their
beliefs in favor of scientific method?

To put a correct view of the universe into people's heads we must first get an incorrect view out. People believe a lot of nonsense about the world of phenomena, nonsense that is a consequence of a wrong way of thinking. . . . *The problem is to get them to reject irrational and supernatural explanations of the world, the demons that exist only in their imaginations, and to accept a social and intellectual apparatus, Science, as the only begetter of truth.*[11] (Italics mine.)

In addition to the unsubstantiated belief in a rational mind that can reject irrational explanations, Lewontin ignores another fundamental aspect of human nature—we also learn through profound emotional experiences that contain no elements of reason. These forms of knowledge aren't ideas that can be assessed, tested, and judged as right or wrong. They aren't "facts"; they are ways of seeing the world that are beyond reason and discussion. We get a better (but personal) sense of the nature of grief from listening to Beethoven's late quartets than from analyzing hypoactive medial frontal areas on functional MRI scans; we sense more about the tragicomedy of life from watching Chaplin's tramp than learning that the sun will eventually burn itself out. Tolstoy's struggle with meaninglessness is itself a profound window onto the human condition. Though not scientific "truths," they contribute to our worldview as much as comprehending string theory. Even if our conclusions are wrong (as is often the case with interpretations of experience), it is what we do, and what gives us comfort. If these experiences trigger a sense of the religious, airtight lines of reasoning won't shake this belief.

How persuasive is this quote from Daniel Dennett, director of the Center for Cognitive Studies and professor of philosophy at Tufts University?

I have absolutely no doubt that the secular and scientific vision is right and deserves to be endorsed by everybody, and as we have seen over the last few thousand years, superstitious and religious doctrines will just have to give way.[12] (Italics mine.)

To insist that the secular and the scientific be universally adopted flies in the face of what neuroscience tells us about different personality traits generating idiosyncratic worldviews. Try telling a poet to give up his musings and become a mechanical engineer. Or counsel a clown that he'd be more useful as a mortician. Perhaps the best example of how basic personality affects perspective, including feelings of meaning and purpose, is the degree to which one has a sense of humor, including a highly developed sense of the ridiculous. For me, Beckett's portrayal of meaningless is both hilarious and curiously uplifting. Watch a good Beckett production and you often find yourself nodding at others in the audience. The great mystery is how the humorous presentation of pointlessness creates its own deep sense of inexpressible meaning, including a feeling of camaraderie with others sharing the same ironic viewpoint.[13]

There's another problem with Dennett's insistence on the absolute correctness of the secular and scientific vision, one that brings us back to the inherent problem of objectivity. In a recent interview in *Salon.com*, Dennett was asked, "Are you saying a person is better served by relinquishing his faith in search of a more rational truth about the universe?" Dennett answered, "That's a very good question and I don't claim to have the answer yet. That's why we have to do the research. Then we'll have a good chance of knowing whether people are better served by reason or faith."[14]

How objective would a study of reason versus faith be if undertaken by someone who has "absolutely no doubt that the secular and scientific vision is right"? And what kind of research could possibly determine whether faith or reason serves us better? This is the same line of reasoning that led the cognitive scientist Ap Dijksterhuis to claim that the best decisions were the ones with which the study participants were the happiest. Should the final arbiter between faith and reason be that which makes us happiest? Or should we choose that which makes us most able to face death? Would a final moment of comforting faith be worth more than a preceding lifetime of reason-based skepticism? I cannot imagine a more unreasonable assumption or a greater leap of faith than believing that you can conduct a scientifically valid research project that will show whether we are better served by reason or faith.

Man is the only creature who refuses to be what he is.

—Albert Camus

In writing this book, I have frequently revisited Darwin's description of his personal struggles with purpose, meaning, and the question of God. In a few paragraphs of his autobiography,[15] he addressed many of the questions at the heart of this book—from the nature and accuracy of profound feelings of conviction to the limits of what we can know. Particularly inspiring is his attempt to accommodate contradictory impulses without either slipping into despair or adopting an absolutist position.

To begin his discussion, Darwin described how the experience of "sublime feelings" while in the midst of the grandeur of a

Brazilian forest led him "to the firm conviction of the existence of God, and of the immortality of the soul." But gradually, over a period of years, such majestic scenes failed to evoke these feelings. He compared this loss of a sense of personal conviction to becoming color-blind in a world that has a universal belief in redness. He was quick to recognize that his lack of conviction—like color blindness—didn't shed any light on any external truth such as whether or not redness exists. "I cannot see that such inward convictions and feelings are of any weight as evidence of what really exists."

He went on to equate the feeling of conviction with other sublime feelings that I have referred to as the *feeling of faith*. "The state of mind which grand scenes formerly excited in me, and which was intimately connected with a belief in God, did not essentially differ from that which is often called the sense of sublimity; and however difficult it may be to explain the genesis of this sense, it can hardly be advanced as an argument for the existence of God, any more than the powerful though vague and similar feelings excited by music."

Darwin was sufficiently astute and introspective to realize that the source of his former belief in God was a mental sensation that had no bearing on any external reality. But he wasn't any easier on the ability of reason to decipher the universe. (The following three paragraphs have been edited for brevity.)

Another source of conviction in the existence of God connected with the reason and not the feelings, impresses me as having much more weight. This follows from the extreme difficulty or rather impossibility of conceiving this immense and wonderful universe, as the result of blind chance or necessity. When thus reflecting, I feel compelled to look at a first cause having an intelligent mind in some degree analogous to that of man; and I deserve to be called a theist.

This conclusion was strong in my mind . . . when I wrote the *Origin of Species;* since that time it has very gradually . . . become weaker. But then arises the doubt—can the mind of man, which has, as I fully believe, been developed from a mind as low as that possessed by the lowest animal, be trusted when it draws such grand conclusions? May not these be the result of the connection between cause and effect which strikes us as a necessary one, but probably depends merely on inherited experience?

I cannot pretend to throw the least light on such abstruse problems. The mystery of the beginning of all things is insoluble to us; and I for one must be content to remain an Agnostic.

Darwin began by admitting that it is impossible to conceptualize the universe as mere blind chance, but ended up accepting that he cannot know when apparent cause-and-effect is nothing more than a trick of the mind. By acknowledging the limits of knowledge based upon both reason and feelings, he unflinchingly accepted that the mind isn't capable of solving the mystery of existence.

Both Darwin and Collins experienced a mystical moment while immersed in nature. For both, the experience was initially profound and seemingly indicated the presence of God. Collins converted from a lifelong atheism to being deeply religious. But Darwin took the opposite tack. A Christian at the time of his trip to the Amazon, he subsequently reinterpreted his feeling of having experienced God as nothing more than a biological trick of his mind, even possibly inherited. Eventually he abandoned Christianity and became an agnostic.

I couldn't ask for a better description of how individual differences in the hidden layer create such dissimilar interpretations of a similar experience. Different genetics, temperaments, and experience led to contrasting worldviews. Reason isn't going to bridge this gap between believers and nonbelievers. Whether an idea originates in a *feeling of faith* or appears to be the result of pure reason, it arises out of a personal hidden layer that we can neither see nor control.

A Practical Suggestion?

Darwin's theory of evolution arose from a self-admitted biased mind, but he honed his biased ideas into a testable hypothesis. After one hundred and fifty years, the evidence confirming the theory of evolution is overwhelming. Nevertheless, we still are saddled with the possible unreliability of consensus opinion as exemplified by the basketball-gorilla video as well as the general issues inherent in objectivity. To solve these problems, Stephen Jay Gould offered this practical compromise: "In science, 'fact' can only mean confirmed to such a degree that it would be perverse to withhold provisional assent."[16]

The key phrase is *provisional assent*. We can strive for objectivity; we cannot reach the shores of dispassionate observation. The problem is that to play according to the rules of scientific method, we must concede the possibility that we cannot know if one day contrary evidence might appear and overthrow a cherished theory. Faith-driven arguments, by invoking irrefutable divine authority that will always be right, do not have to make this concession. This uneven playing field isn't going to go away. The problem becomes particularly acute when evolution is seriously questioned by nearly half of Americans: "In a 2001 Gallup poll 45 percent of U.S. adults said they believe evolution has played no role in shaping humans. According to the creationist view, God produced humans fully formed, with no previous related species."[17]

The choices for how to proceed aren't particularly satisfying. To admit that evolution should only be granted provisional assent is to concede that an alternative explanation—creationism or intelligent design—might possibly be right. To elevate evolution to an unequivocal fact is to perpetuate the biologically unsound myth of the autonomous rational mind that provides faith-driven arguments their best tool for confirmation—the one-step checkmate of "I know what I know."

If science is to carry on a meaningful dialogue with religion, it must work to establish a level playing field where both sides honestly address what we can and cannot know about ourselves and the world around us. We need to back away from perpetuating the all-knowing rational mind myth that makes real discussion impossible. At the same time, we need to acknowledge that the evidence for a visceral need for a sense of faith, purpose, and meaning is as powerful as the evidence for evolution. And we must factor in that irrational beliefs can have real adaptive benefits—from the placebo effect to a sense of hope. Insistence upon objectivity and reason should be seen within a larger picture of our biological needs and constraints.

The goal of this dialogue should be to maximize personal hope and a sense of meaning while minimizing the untoward effects of unjustifiable personal attitudes and social policies. We should force ourselves to distinguish between separate physiological categories of faith—the basic visceral drive for meaning that has real purpose versus the unsubstantiated cognitive acceptance of an idea. Compassion, empathy, and humility can only arise out of recognizing that our common desires are differently expressed.

If possible, both science and religion should try to adopt and stick with the idea of provisional facts. Once all facts become works-in-progress, absolutism would be dethroned. No matter how great the "evidence," the literal interpretation of the Bible or Koran would no longer be the only possibility. By exploring and making common knowledge of how the brain balances off contradictory aspects of its biology, we might gradually turn absolutism into an untenable stance of ignorance.

We don't ask that people poke out their eyes to prevent them from making mistaken eyewitness identifications; instead, we demonstrate the power of perceptual mischief via optical illusions and courses in perceptual psychology. Imagine how different dialogue might be with future generations raised on the idea that there are biological constraints on our ability to know what we know. To me, that is our only hope.

14

Mind Speculations

REALIZING THAT WE KNOW OUR THOUGHTS THROUGH MEN-
tal sensations that are subject to perceptual illusions and mis-
perceptions has prompted me to wonder if some of the toughest
age-old philosophical issues arise out of attempts to resolve perceptual tricks created by our brains. This chapter is not meant to
be a "one theory fits all problems" tidying-up. Nevertheless, I'd like
to spend a few pages thinking about how some of the greatest
metaphysical puzzles might be nothing more than unavoidable
by-products of conflicting biology.

A classic example is the optical illusion of the silhouette of two
opposing faces that can also be seen as a vase. Stare at the picture
and the vase will alternate with the facial profiles. You cannot will
yourself to continuously see either the faces or the vase. This unstable alternating relationship of foreground to background is the
result of a perpetual tug-of-war between equally weighted aspects
of visual perception. The question that we ask ourselves—which is

it, two faces in silhouette or a vase?—has no answer even if it feels as though it might. The question has no real meaning; it is nothing more than an attempt by the hidden layer to resolve competing aspects of perception. It might be said that the problem of deciding between faces and a vase doesn't exist outside of the mind of the viewer. It isn't a "real world" issue. Consider this foreground-background tug-of-war as a model of a biologically generated paradox that cannot be resolved.

With such impersonal visual tricks, we can shrug off the resulting lack of resolution by telling ourselves that this is an optical illusion. Though we don't feel a sense of satisfaction of having the image come to rest, knowing why this isn't possible prevents us from feeling compelled to choose definitively between faces or a vase. We remind ourselves that this is pure biology on display, and move on to other thoughts. But with unstable mental images of ideas that are personally meaningful, this is far more difficult.

The Origin of the Universe or Cosmology Versus Edges and Borders

In the beginning there was nothing at all except darkness. All was darkness and emptiness. For a long, long while, the darkness gathered until it became a great mass.

—Arizona Pima Indian oral story

My car's recently replaced rear bumper is sad proof that a gray utility pole isn't readily visible on a foggy night. My excuse is that I was a victim of basic neurophysiology. We determine shapes by seeing borders; it is impossible to clearly see an object without a sharply contrasting background. The same optics govern our mind's eye. Close your eyes and try envisioning a face. You will see it against some kind of contrasting background, whether it is a neutral color or a vague grayness or blackness. Now try to visualize a perfect vacuum. Even if I know that a vacuum contains nothing, there is still an "it," a nothingness that must exist within some type of space. My mind serves up a dim empty darkness as it simultaneously tells me that this can't be so.[1] Empty space is a visual non sequitur; there is no visual counterpart of nothingness.

Let's move on to cosmology. Try to visualize the big bang—a single infinitely dense point that suddenly explodes. To see this object in our mind's eye, we place this dot against some contrasting background. Most people, when questioned, will offer that they see a dim darkness against which the initial singularity is framed. This problem of borders isn't confined to spatial considerations; time is equally impossible to visualize as either always existing or suddenly beginning. We see a beginning in contrast to what was present *just before* the beginning. The cruel irony is that a mind's eye's representation of no surrounding space or time occupies some space and suggests a prior time. To relieve the resulting tensions, we feel compelled to ask a key question shared by science and religion—what, if anything, was present before the beginning?

I have tried to imagine how this question might be framed if we had a different visual apparatus that didn't require an object to be seen against a background. But I am stuck with the limits of my mind's eye in the same way that my brain cannot resolve the vase-faces illusion. Whether the question would even exist if we had a different mind's eye isn't answerable. We can't know if a question such as what came before the beginning has any more meaning than trying to decide whether we are looking at faces or a vase. Even sensing that the question is "real" isn't evidence, as we've repeatedly seen in the chapters on the involuntary nature of the feeling of "realness." (This is also an example of how reason cannot be separated from bodily sensations. Any notion of space—no matter how abstract—must be filtered through our bodily perceptions of space. In our mind's eye, emptiness occupies space.)

How we approach this problem will be influenced by prevailing cultural attitudes. If we are told that the vase-faces picture is definitely either two faces or a vase and that we must make a choice, we will spend considerable time trying to arbitrarily pick one image over the other. We will then work on convincing ourselves that this answer is correct. Some will remain skeptical; others will become convinced. This conviction is the equivalent of blind unsubstantiated faith. But if we are told that the inability to choose one over the other is a function of how our brains work, we would be more likely to accept that the illusion can't be resolved. (We don't expect a glass rod to appear straight when it is half-immersed in a glass beaker because we have learned the laws of refraction.)

If scientifically inclined, we gravitate toward theories of universe upon universe, or universe before universe, to possibly correct, but unfathomable mathematical equations that show how the universe can enfold itself without requiring a surrounding space. But none seem capable of resolving this inner mental tension. The following description from Nova's *History of the Universe* creates more unanswered questions than it solves: "The universe began with a vast explosion that generated space and time and created all the matter in the universe."[2] The *Scientific American* explanation is equally unsatisfying. "The point-universe was not an object isolated in space; it was the entire universe, and so the only answer can be that the big bang happened everywhere."[3]

Even the most brilliant are not immune. Stephen Hawking has said, "The idea that space and time may form a closed surface without boundary . . . has profound implications for the role of God in the affairs of the universe. . . . So long as the universe had a beginning, we could suppose it had a creator. But if the universe is really completely self-contained, having no boundary or edge, it would have neither beginning nor end. What place, then, for a creator?"[4] In an attempt to circumnavigate this mind's eye boundary issue, Hawking has postulated a "no-boundary" state—an idea that, even if entirely correct, isn't consistent with how our mind's eye works. We want a palpable resolution for the tension created by trying to understand the surrounding background, not an abstraction that we can't see or feel.

If science can't provide resolution, most will look elsewhere—from theories of a creator existing prior to the origin of the universe to an intelligent design that brought the universe into existence. To put this physiological dilemma into a cultural and historical perspective, a quick Google search reveals more than five hundred different creation myths. (Note the similarity between the Pima Indian creation myth and the big bang hypothesis.) As long as those in power—both scientists and religious leaders—insist that we can know how the universe came into existence, we will be similarly tempted. Hypotheses ranging from a grand creator to intelligent design to a no-boundary universe are the inevitable consequences of believing in answers even when the questions may reflect nothing more than quirks of brain physiology.

Of historical note is that more than two hundred years ago, Immanuel Kant proposed that the physical mechanisms that shape the perceptions of our experiences also shape the way we think about those phenomena that we cannot directly experience. At a time when the brain was a mysterious organ and neuroscience wasn't even science fiction, Kant anticipated the discovery of brain functions that can both influence and even generate major philosophical concerns.[5]

IN A 1991 short story by Terry Bisson, a robotic commander of an interplanetary expedition reported to his electronic leader that the human inhabitants of Earth are "made out of meat."

"Meat?"

"There's no doubt about it."

"That's impossible. . . . How can meat make a machine? You're asking me to believe in sentient meat."

"I'm not asking you. I'm telling you. These creatures are the only sentient race in the sector, and they're made out of meat."

"Spare me. Okay, maybe they're only part meat. . . ."

"Nope, we thought of that, since they do have meat heads. . . . But . . . they're meat all the way through."

"No brain?"

"Oh, there is a brain all right. It's just that the brain is made out of meat!"

"So . . . what does the thinking?"

"You're not understanding, are you? The brain does the thinking. The meat."

"Thinking meat! You're asking me to believe in thinking meat?"

"Yes, thinking meat! Conscious meat! Dreaming meat! The meat is the whole deal! Are you getting the picture?"[6]

Conscious, thinking, dreaming meat—this powerful image of mindless flesh producing our most prized traits serves as an apt introduction to the age-old question: Is the mind separate from the machinery that creates it? Rather than jumping into the debate, I'd suggest that we first look to see if the mind-body dualism issue—like the faces-vase illusion—is nothing more than the interplay of contradictory biological forces.

Mind-Body Dualism and the Sense of Self

Pain's purpose is to tell us when some part of our machinery has gone awry. Hunger and thirst tell us when we need to refuel and drink up. To be meaningful, these sensations must feel as though they reflect the underlying physical status of our bodies. But other sensations serve us best when they are divorced from any awareness of bodily functions. The most immediate example is the *sense of self*. At the risk of falling into the "everything has an evolutionary explanation" trap, it is easy to speculate that an individual sense of self was instrumental in the development of morality, compassion, laws, goals, higher purpose, and meaning—all the various prerequisites for social order. Essential to this perception of being a unique and valuable individual is not feeling that the self is just the product of underlying "mindless" neurons.

We readily acknowledge that pain is a purely subjective sensation that emerges from pain receptors and pain-generating mechanisms within the midbrain and thalamus. It has no substance or weight; we cannot send it to the lab for anatomic analysis. Like all mental states, it doesn't exist on its own, but rather is an extension of underlying biological mechanisms. Arbitrary categories such as mental versus physical are woefully inadequate to describe this complex interaction of real neurons and synapses and exclusively subjective mental states. Nevertheless, we aren't particularly bothered with philosophical questions about the existence of pain; we accept that the pain of a stubbed toe is "real" even though devoid of any physical properties normally associated with "realness."[7]

Nor are we surprised when this purely subjective sensation isn't always located "where it should be." For a moment, consider the problem of referred pain. From an evolutionary standpoint, a pain warning system should accurately localize potential problems. The pain from a stubbed toe should immediately draw your attention to your toe, not your elbow. But biology sometimes leads us astray. For example, you are running uphill on a cold day and get a deep aching in your left arm. When you stop, the pain subsides. You check your arm and there's nothing wrong; it moves freely and painlessly. You start jogging again and the pain returns. How you interpret the pain depends upon your education, experience, and age. Without needing to understand the underlying physiology, most of us of a certain age would immediately worry that the problem was cardiac—a coronary artery insufficiency.

The explanation is quite straightforward. The heart and left arm both originate from the same region of the developing embryo. Sensory inputs from both the arm and the heart are processed in the same segments of the spinal cord. If there is an overflow of the incoming pain fiber impulses, they can be felt in other areas served by the same region, causing the false localization of *referred pain*. Thanks to radio and TV public service messages, we realize that pain in the left arm can be a warning symptom of a heart attack. Rather than ruminate on the philosophical implications of whether or not the exercise-induced left arm pain is "really there," we call 911 or head for the nearest emergency room. The point is that we can learn how to cope with perceptual misdirections without feeling obliged to drum up far-fetched metaphysical explanations. We are satisfied that the left arm pain—although not detectable or measurable—is a very real signal from a normally functioning pain warning system.

The same understanding and categorization should apply to the sense of self—another subjective mental state that arises from neurons and synapses. But we have a problem. Pain feels as though it is a reflection of the underlying physical state of our body; the sense of self doesn't. The two emergent phenomena—pain and a sense of self—have very different agendas. One is to point to the body and give warning signals. The other is to point away from the body in order to create a sense of individuality above and beyond mere biology. To have any sense of personal meaning, we must see ourselves as more than mere machinery or thinking meat. This separate sense of self, like the perceived dark emptiness that surrounds the big bang at the beginning of the universe, feels as though it requires an explanation for its independent existence. The result is the cognitive dissonance of intellectually knowing that the brain must create the sense of self versus the necessary feeling that the self is separate from the brain. At a physiological level, this isn't fundamentally different than a patient with Cotard's syndrome feeling her beating pulse yet still believing that she is dead. To put this into perspective, see how you feel about the following statement by the contemporary philosopher, John Searle: "Conscious states are entirely caused by lower level neurobiological processes in the brain. . . . They have absolutely no life of their own, independent of the neurobiology."[8]

The centuries-old Cartesian mind-body dualism issue hinges on how you perceive the above passage. The statement has an extremely high likelihood of being correct, yet it is hard to imagine how to read this statement without feeling a separate you doing the reading, experiencing the pleasure of immediately knowing the underlying neurophysiology, or feeling a sense of pride in finding confirmation in something you already suspected. But what is the point of feeling pride in knowing that we are mere machinery? We aren't proud of being thirsty or having our food properly digested. What would be the pleasure of understanding if it were not somehow a reflection on our overall character, intelligence, wisdom, or sophistication . . . ? What would be the purpose of having this not immediately practical knowledge if it weren't to enhance one's sense of self?

It is only by having conscious states that feel independent of their biology that we can understand what the passage means. Trapped within our biology, we cannot escape the mind-body dualism issue. It is part of who we are. A full exposition of the underlying brain mechanisms won't prevent us from seeking larger meanings any more than understanding the big bang theory stops us from wondering what surrounds the universe or came before the beginning. It is our fate. I cannot imagine an existence in which we didn't ponder our existence, including who we are collectively and individually. The alternative—that we are just bags of chemicals—will never be a bestseller.

All theory is against the freedom of the will; all experience is for it.

—Samuel Johnson, quoted in *Boswell's Life of Johnson*

To offer a final example of how mental sensations can create philosophical conundrums, let's conclude this section with a brief look at free will. Imagine a two-year-old who finds any noise bothersome and tells his parents not to play the TV, radio, or stereo when he's in the house. If Michael Merzenich's studies are correct, the two-year-old's choice to make the house church-quiet will affect the future development of his auditory cortex. His seemingly voluntary and intentional mental decision will create permanent physical changes in his brain. If the two-year-old were a nascent philosopher, he would offer himself as living proof of the interface between free will and "hardwiring."

On the other hand, there is an extensive but controversial body of neuroscience literature claiming that such choices are made in the unconscious *prior* to the child consciously sensing that he's made the choice. (The studies of Ben Libet are central to this issue, and are well outlined in his book *Mind Time*.)[9] The argument is that unconscious thoughts trigger our behavior and our conscious explanations follow at a distance.[10]

But we've already seen the problem with defining intentional and willful. In the *Pogo* example, the sudden appearance of the answer didn't feel willful, but the unconscious was provided with a clear assignment: To remember the possum's name. My decision whether or not the mental process that came up with *Pogo* was intentional is based on how the answer *feels*, not a basic understanding of what went on in my unconscious. Choice without the feeling of choice is, "It just occurred to me." Choice with the feeling of choice is, "Yes, that is my final decision." The feeling of choice is a poor indicator of underlying intent.

Rapid motor movements offer the same problem. As Ben Libet puts it, "The playing of a musical instrument, like the piano, must involve an unconscious performance of the actions. Pianists often play rapid musical runs in which the fingers of both hands are hitting the keys in sequences so fast that they can barely be followed visually. Not only that, each finger must hit the correct piano key in each sequence. Performers report that they are not aware of the intention to activate each finger. Instead, they tend to focus their attention on expressing their musical feelings. Even these feelings arise unconsciously, before any awareness of them develops."[11]

To say that a pianist has no awareness of an intention to strike each key in sequence doesn't mean that he found himself playing at Carnegie Hall quite by accident or because of the whim of the gods of fate. The performance is quite intentional. What is lacking is the pianist's awareness of this sense of intention as he is playing. This isn't surprising; conscious perception of an intent to hit a particular note takes longer than the motor response to play the note. (The musical equivalent of the approaching baseball example.) During this perceptual delay, the pianist will have played a flurry of subsequent notes. Being aware of an intention to hit notes already played wouldn't make sense, and would slow us down to the level of our first piano lessons, when every note was struck after conscious deliberation. Suppressing any feeling of intention is a necessary prerequisite for rapid motor movements.

In both the *Pogo* and piano-playing examples, a lack of sense of intention tells us nothing about underlying intention. Ironically, how we feel about the willfulness of a choice is beyond our control. To flesh out this diabolical paradox, briefly consider Tourette's syndrome.

In 1965, at a University of California, San Francisco, pediatric neurology conference, the patient, a frightened fifteen-year-old Asian boy, was being interviewed by the neurology department chairman, a tall, stately man in a monogrammed white lab coat.

"No, sir," he said. "I have no idea why I have these tics." He stared at his feet.

"Come on," the chairman insisted. "Surely you have some explanation."

The boy shrugged, the movement expanding into a series of head jerks, eye blinks, and lip smacking. "No, sir," he blurted out, eyes still averted.

"You mean that you grunt and grimace for no reason at all?" The chairman scowled.

The boy's tics accelerated. The boy stood his ground, pinching his lips together, fighting an urge we all knew by history to be the boy's chief complaint.

"No reason at all?" the chairman repeated. "Everything is just ducky?" The chairman turned to his neurological brethren crowded into the overheated conference room, half-smiling under his half-glasses. Soon, every neurologist thought. Soon.

The boy looked from the chairman to the audience, then back to the chairman again. "Duck, duck, fuck a duck, fuck a duck, fuck a doc, fuck you, doc. . . ."

The chairman beamed, pleased with his clinical astuteness in instigating the outburst.

"Fuck you, doc," the boy continued, unable to control himself.

The chairman's smile collapsed; his face was bright red. "Stop that," he said, grabbing the boy by the shoulder. The boy couldn't stop. The chairman exploded. "Transfer him to Langley Porter [the psychiatric facility attached to the University of California hospital]. Maybe they can teach him some manners."

I watched as the humiliated youth was ushered out of the conference room. Something was dreadfully wrong. The chairman was a seasoned clinician and had intentionally provoked the boy. He had known what to expect, yet had taken the outburst personally. But if the cursing was reflexive, no more than a pathological knee-jerk response arising from a neurological malfunction, it couldn't be personal. (Neurological colleagues from other countries tell me that coprolalia—the difficult-to-control scatological outbursts seen in a small fraction of patients with Tourette's syndrome—are fairly uniform, but specific to the vernacular of each country.)

More than forty years later we neurologists offer that Tourette's syndrome is a predominantly genetic disorder with the prime suspect being faulty brain neurotransmitter metabolism, predominantly dopamine. On the surface, most of us have made a conceptual about-face and are willing to accept that the uncontrollable foul language of coprolalia results from disordered neurochemistry, not a twisted psyche.

From a Tourette's support group we learn that:

The coprolalia type outburst usually disrupts communication, speech, or something that a patient is involved in. Following the disruption, the patient continues about their communication, speech, or project normally. These disruptions will usually continue to enter in and out of a patient's normal behaviors and events.

For example, a patient with coprolalia could be talking with someone who mentions the word *duck*. The word *duck* trips a vocal tic in the coprolalia patient of which follows three quick vocal bursts of, "Fuck a duck, fuck a duck, fuck a duck." The conversation keeps flowing as it was prior to the vocal disruption.

An observer, who is not familiar with coprolalia nor understands it, may believe the outburst is the result of a conscious and voluntary decision to swear. *However, the outbursts are neither intentional nor purposeful.*[12] (Italics mine.)

Now listen to this description from a patient with Tourette's syndrome:

> I effectively *never* swear . . . but in times of great stress the profanities just keep coming! Like all tics, it starts as a nagging itch that something is wrong. But instead of moving to scratch it, you have to say words. *I choose the ones that would be most offensive* to whoever is nearby. You tend to use the ones you consider to be the worst. Then, of course, when I've relaxed a bit, the memory of what I've said haunts me for ages.[13] (Italics mine.)

How are we to reconcile these two very different claims? The patient feels that he can consciously and deliberately choose which words to use, yet the very essence of a tic is an involuntary and meaningless motor movement or vocalization. So, is the patient's sense of choice of words real or an illusion, or is he capable of knowing the difference? Is this an example of the brain giving the patient a false sense of choice in order to avoid the more frightening acknowledgment that he isn't in control of his mind? Or are we to postulate the even more confusing proposition that he willfully selected which words he would involuntarily utter?

We could endlessly speculate, but before we can seriously address issues like free will, we need to ask the more basic question: What exactly is a *sense of choice?* What are the basic brain control mechanisms that determine when the feeling is present along with a cognitive choice (my choosing to write this sentence), or absent despite full intention (the *Pogo* and piano-playing examples), or present in the absence of apparent choice (as in the Tourette's patient)?

Which leads to a more practical issue—the nature of personal responsibility. After reading the Tourette's patient's explanation, do you feel that he is completely, partially, or not at all responsible for his explosive outbursts of swear words? And how would you decide? Is there a single optimal line of reasoning that we should all adopt? Do you feel that you can consciously and willfully make this decision? Much of this book has been devoted to showing how thought arises out of a hidden layer filled with innate bias. We have seen where genetic predispositions influence our thoughts. How are we to think about personal responsibility that arises out of such a messy and ill-defined cognitive stew?

Imagine having a close friend cheat you in a business deal. You want to get even, but you tell yourself to, "Get over it." Twenty years pass; you don't see the former friend or consciously give him a second thought. Then one day you bump into him on the street. He acts as if nothing had ever happened. You are infuriated and blurt out that he's a no good rotten scoundrel. He shrugs, laughs mockingly, and makes you feel ridiculous. Suddenly, without apparent thought, you push him backward. He slips, falls, and breaks his shoulder. You are charged with assault and battery, and are personally sued for damages. Your defense is "I didn't mean to push him. The thought never entered my mind. I don't know what came over me. I wasn't myself."

But if your unconscious could speak, it would disagree. It would tell you that it was only acting on your twenty-year-old desire. The irony is that the most involuntary-appearing act may arise out of a stored intention of which you have no knowledge. If free will implies the ability to make choices, then your unconscious will have made a choice that you don't consider a choice.

When contemplating the degree of personal responsibility each of us has for his actions, we are immediately up against the constraints of how we experience ourselves. Mental sensations will prompt us to feel or not feel that we are choosing and that we can know when such "thoughts" are correct. Combine a *feeling of knowing* with a *feeling of choice* and you can begin to see the immense complexity of "knowing when you have made a willful choice." Just as mental and physical are arbitrary classifications that cannot adequately describe emergent phenomena, the free will–determinist debate is limited by its own biological constraints.

WHETHER THINKING ABOUT the origins of the universe, the presence or absence of a soul, or deciding on free will and personal responsibility, we need to step back and first consider how these problems are influenced by a variety of mental states over which we have no conscious control. Mental sensations are the cornerstones of thought. Before we can address the great philosophical questions, we need to know how these questions are themselves the product of our biology, and in particular the various mental sensations that give our thoughts felt meaning.

A personal digression: Since starting this book, I increasingly find myself asking myself a rhetorical question, "How would an alien from Mars approach this issue?" For example, take the origin of the universe question. What if he had a silicon-based brain that operated without a mind's eye visualization? How would the problem of borders and boundaries be tackled, or would the problem even exist? Of course, I cannot imagine this, but I can imagine the possibility. That is enough to keep me from slipping into absolutes. Trying to pose the question from an alternative biological perspective forces me to quickly acknowledge the limits of my own thoughts.

15

Final Thoughts

Not ignorance, but ignorance of ignorance, is the death of knowledge.

—Alfred North Whitehead

There must be certainty from the U.S. president.

—George W. Bush

A Brief Recap

THE FEELINGS OF KNOWING, FAMILIARITY, STRANGENESS, and realness are more than neurological curiosities associated with complex partial seizures and temporal lobe brain stimulations. And they don't fit neatly into standard categories of mental functions—emotions, moods, or thoughts. Collectively they represent aspects of a separate type of mental activity: an internal monitoring system that makes us aware of and colors, judges, and assesses our thoughts.

The most obvious analogy is to the body's various sensory systems. It is through sight and sound that we are in contact with the world around us. Similarly, we have extensive sensory functions for assessing our interior milieu. When our body needs food, we feel hunger. When we are dehydrated and require water, we feel thirsty. If we have sensory systems to connect us with the outside world, and sensory systems to notify us of our internal bodily needs, it seems reasonable that we would also have a sensory system to tell us what our minds are doing. To be aware of thinking, we need a sensation that tells us that we are thinking. To reward learning, we need feelings of being on the right track, or of being correct. And there must be similar feelings to reward and encourage the as yet unproven thoughts—the idle speculations and musings that will eventually become useful new ideas.

To be effective powerful rewards, some of these sensations such as the *feeling of knowing* and the *feeling of conviction* must feel like conscious and deliberate conclusions. As a result, the brain has developed a constellation of mental sensations that feel like thoughts but aren't.

These involuntary and uncontrollable feelings *are* the mind's sensations; as sensations they are subject to a wide variety of perceptual illusions common to all sensory systems. For example, temporal alterations in the experience of time are everyday occurrences in the visual system (the approaching baseball example). Applying this understanding to mental sensations can help us see that the *feeling of knowing* might seem as though it is occurring in *response* to a thought when it actually preceded the thought and was responsible for bringing the thought into awareness (the "This must be Izzy Nutz's house" example).

Appreciation that the brain's hierarchical structure is organized along the general lines of neural networks also allows us to see this mental sensory system as integral to the formation of a thought. Earlier, in chapter 5, I described each neural network within a larger neural network as being analogous to one member of a larger committee. A question is posed (input). Each committee member has a single vote; once all the votes are tallied (the hidden layer calculation), a final decision is made (output). Now imagine a neural network in which each committee member represents one of the mental sensations—from a *feeling of knowing* to a feeling of familiar, bizarre, or real. It will be the final tally of votes that will determine how we feel about a thought, including its "rightness" or "wrongness." Before reading the answer, the committee members of the neural network evaluating the description of a kite paragraph will vote for unfamiliar, strange, perhaps even bizarre or unreal. There will be no votes for a sense of understanding. When the explanation—kite—is inputted, the committee members for familiar "yes, that's correct" and a *feeling of conviction* will override the suddenly silent committee members representing strange and unfamiliar. The final result is that the explanation will *feel correct*.

Once imbedded within the conclusion that this paragraph refers to a kite, the *feeling of correctness* cannot be consciously dislodged or diminished. We can consciously input new contrary information; only the hidden layer of the neural networks can reweight the values.

The message at the heart of this book is that the *feelings of knowing, correctness, conviction,* and *certainty* aren't deliberate conclusions and conscious choices. They are mental sensations that *happen* to us.

Some Ideas Are More Equal Than Others

We laugh at a magic trick, and develop theorems to explain why a glass rod half-immersed in water appears bent. We cannot train ourselves to see the sleight of hand that makes it impossible to

win at three-card monte, but we can tell ourselves that we are being deceived and not to trust what we see. Let this be the model
for the *feeling of knowing*. Neuroscience needs to address the physiology; we need to question the feeling. And nothing could be
more basic than to simply question the phrase, "I know."

As we've seen, the standard definitions of *to know*—to perceive
directly; grasp in the mind with clarity or certainty; to regard as
true beyond doubt—are inconsistent with our present-day understanding of brain function. Somehow we must incorporate what
neuroscience is telling us about the limits of knowing into our
everyday lives. Imagine applying this simple principle to the *Challenger* study. Instead of saying, "That's my journal and my handwriting, but that's not what happened," the students might learn
to say, "That's my journal entry, but it doesn't *feel right* anymore."
Perhaps the easiest solution would be to substitute the word *believe* for *know*. A physician faced with an unsubstantiated gut feeling might say, "I believe there's an effect despite the lack of
evidence," not, "I'm sure there's an effect." And yes, scientists would
be better served by saying, "I believe that evolution is correct because of the overwhelming evidence."

I realize that this last sentence runs against the grain of those
who have fought the hardest to establish science as the method
for determining the facts of the external world. It is particularly
loathsome when you feel that you are playing into the hands of
religious fanatics, medical quacks, and word-twisting politicians.
But substituting *believe* for *know* doesn't negate scientific knowledge; it only shifts a hard-earned fact from being unequivocal to
highly likely. To say that evolution is extremely likely rather than
absolutely certain doesn't reduce the strength of its argument, at
the same time as it serves a more fundamental purpose. Hearing

myself saying, "I believe," where formerly I would have said, "I know," serves as a constant reminder of the limits of knowledge and objectivity. At the same time as I am forced to consider the possibility that contrary opinions might have a grain of truth, I am provided with the perfect rebuttal for those who claim that they "know that they are right." It is in the leap from 99.99999 percent likely to a 100 percent guarantee that we give up tolerance for conflicting opinions, and provide the basis for the fundamentalist's claim to pure and certain knowledge.

A related consideration is to distinguish between felt knowledge—such as hunches and gut feelings—and knowledge that arises out of empiric testing. Any idea that either hasn't been or isn't capable of being independently tested should be considered a personal vision. Shakespeare does not demand that we accept *Hamlet* as representing a universal truth. We agree and judge him according to the standards of art, literature, and personal experience. *Hamlet* is neither right nor wrong. If in the future, Hamlet is found to have a gene for bipolar disorder, we are entitled to reassess our initial interpretations of Hamlet's relationship to his mother. Hamlet is a vision. So are each of the quotes cited in these last chapters. No matter how seemingly reasonable and persuasive, each begins with a very idiosyncratic perception that seeks its own reflection in the external world. Each writer's personal sense of purpose drives the arguments, picks out the evidence, and draws conclusions. Such ideas should be judged accordingly—as visions, not as obligatory lines of reasoning that must be universally shared.

To retreat from claims of absolute "knowing" and certainty, popular psychology needs to explore how mental sensations play a fundamental role in generating and shaping our thoughts. We can't afford to continue with the outdated claims of a perfectly rational unconscious or knowing when we can trust gut feelings. We need to rethink the very nature of a thought, including the recognition of how various perceptual limitations are inevitable.

At the same time, if the goal of science is to gradually overcome deeply embedded superstition, it must be seen as a more attractive and comforting alternative, not as inflammatory exhortation and confrontation with a none-too-subtle whiff of condescension. Try to peddle the vision of a cold, pointless world at a Pentecostal revival meeting and you have an inkling of the challenge. In a recent survey, nearly 90 percent of Americans expressed the belief that their souls will survive the death of their bodies and ascend to heaven.[1] Such beliefs, no matter how counter to the evidence, provide the majority of Americans with a personal sense of meaning. If forced to choose between reason and a sense of purpose, most of us would side with purpose. As we've seen, this apparent choice isn't even an entirely conscious decision. If science hasn't yet made a dent in such beliefs, it seems unlikely that further efforts will miraculously turn the tide.

Such discussions pose the same ethical problems inherent in placebo treatments. Simply put, a placebo effect is a false belief that has real value. To insist that there is no soul or afterlife is the moral equivalent of taking away the placebo effect arising out of an unscientific belief. Mr. A's sham arthroscopic surgery allowed him to walk comfortably again. No one should recommend sham knee surgery; the potential downside is too great. Yet many physicians are comfortable recommending less drastic but unproven treatments for pain.

The answer is rarely black or white. Even if the treatment has no risks or cost, the precedent of falsely representing benefits of a treatment has its own long-term undesirable effects. The most serious would be the erosion of trust between the physician and patient. On the other hand, eliminating all placebo treatments because they are intellectually dishonest raises its own set of problems, including the cynical zeitgeist of valuing science over compassion. There isn't easy solution or right answer; each of us will calculate the risk versus reward according to our own biology and experience.

In medicine, we are increasingly developing ethical standards for complex medical decisions that both allow for hope and placebo effect, yet don't fly in the face of evidence-based medical knowledge. The guiding principle of the Hippocratic oath is *primum no nocerum*: Above all, do no harm. This same principle should be a cornerstone of how science competes in the world of ideas. Science needs to maintain its integrity at the same time as it must retain compassionate respect for aspects of human nature that aren't "reasonable."

This balance of opposites extends to all aspects of modern thought. For example, it doesn't make sense to ask someone if he'd like to take a placebo; the very question strips the placebo of much of its intended benefit. Similarly, it isn't clear how to have a reasonable discussion on the nature of the self that both retains the integrity of science—the self is an emergent phenomenon and not some separately existing entity, yet allows each of us to feel that we are individuals and not mere machinery. I cannot imagine a world in which we fully accepted and felt that we were nothing more than fictional narratives arising out of "mindless" neurons. And I cannot imagine how much empathy we would have with others if we saw disappointment, love, and grief solely as chemical reactions. Faced with this chilling interpretation of our lives, it isn't surprising that most people opt for the belief in material "souls" and/or anticipate that real live virgins are patiently awaiting their arrival in heaven.

The Juggling Act

In *The Crack-Up*, F. Scott Fitzgerald described an easy-to-accept but difficult-to-accomplish solution: "The test of a first-rate intelligence is the ability to hold two opposed ideas in the mind at the same time and still retain the ability to function." This is the only practical alternative to cognitive dissonance, where one set of values overrides otherwise convincing contrary evidence. This juggling act requires us to keep in mind what science is telling us about ourselves while acknowledging the positive benefits of non-scientific and/or unreasonable beliefs. Each position has its own risks and rewards; both need to be considered and balanced within the overarching mandate: Above all, do no harm.

Just as we learn to cope with the anxieties of sickness and death, we must learn to tolerate contradictory aspects of our biology. Our minds have their own agendas. We can intervene through greater understanding of what we can and cannot control, by knowing where potential deceptions lurk, and by a willingness to accept that our knowledge of the world around us is limited by fundamental conflicts in how our minds work.

Which leads us back to the central theme of this book. Certainty is not biologically possible. We must learn (and teach our children) to tolerate the unpleasantness of uncertainty. Science has given us the language and tools of probabilities. We have methods for analyzing and ranking opinion according to their likelihood of correctness. That is enough. We do not need and cannot afford the catastrophes born out of a belief in certainty. As David Gross, Ph.D., and the 2004 recipient of the Nobel Prize in physics, said, "The most important product of knowledge is ignorance."[2]

If this book has provoked you to ask the most basic of questions—how do you know what you know?—it will have served its purpose.

Notes

Preface

1. The Phineas Gage information page is maintained by Malcolm Macmillan, School of Psychology at Deakin University, Victoria, Australia, www.deakin.edu.au/hbs/GAGEPAGE.

Chapter 2: How Do We Know What We Know?

1. Neurologically injured patients with an inability to form long-term memories can learn new tasks (such as games, or musical tunes) without any awareness of having previously performed the tasks. With such procedural memory, the patients remember without knowing that they've remembered. Ambulatory patients with advanced Alzheimer's disease can still play golf; their implicit motor skills remain long after they have forgotten their handicap. For an excellent concise categorization of memory see, Budson, A. E., and Price, B., "Memory Dysfunction," *New England Journal of Medicine*, 352, no. 7 (2005). Weiskrantz, L., *Blindsight* (Oxford: Oxford University Press, 1990), is a valuable monograph by one of the pioneer investigators of the phenomenon. Stoerig, P., "Varieties of Vision: From Blind Responses to Conscious Recognition," *Trends in Neuroscience*, 19 (1996): 401–6, provides an in-depth discussion of blindsight as one of several demonstrable dissociations in human visual processing.

2. Neisser, U., and Harsch, N., "Phantom Flashbulbs: False Recollections of
   Hearing the News About *Challenger*," in *Affect and Accuracy in Recall:
   Studies of "Flashbulb" Memories*, Winograd, E., and Neisser, U., eds.
   (New York: Cambridge University Press, 1992): 9–31. In Neisser and
   Harsch's test of the students' accuracy of their subsequent recollections
   of the *Challenger* explosion, a perfect score was 7. In the students tested,
   the mean score was 2.95. Less than 10 percent got a perfect 7, and over
   half got less than 2.

3. Festinger, L., *A Theory of Cognitive Dissonance* (Stanford: Stanford University, 1957).

4. Festinger, L., Riecken, H., and Schachter, S., *When Prophecy Fails* (Min-
   neapolis: University of Minnesota Press, 1956).

5. Weiss, K., *In Six Days: Why 50 Scientists Choose to Believe in Creation*
   (Australia: New Holland Publishers, 1999). A fascinating summary of
   Wise's conversion to creationism is provided in a Richard Dawkins com-
   mentary at www.beliefnet.com/story/203/story_20334_2.html.

6. Moseley, B., et al., "A Controlled Trial of Arthroscopic Surgery for Osteoarthritis of the Knee," *New England Journal of Medicine*, 347, no. 2
   (2002): 81–88.

7. Talbot, M., "The Placebo Prescription," *The New York Times*, January 9,
   2000. Also available at www.nytimes.com.

8. A fascinating overview of misidentification syndromes is provided in
   Hirstein, W., *Brain Fiction: Self-Deception and the Riddle of Confabulation*
   (Cambridge, Mass.: MIT Press, 2005).

Chapter 3: Conviction Isn't a Choice

1. James, W., *The Varieties of Religious Experience* (New York: New American
   Library, 1958), 295.

2. Ibid., 292–93.

3. In *The Varieties of Religious Experience*, James quotes Walt Whitman's
   extraordinary description of a mystical state of "knowing" in the absence of any conscious reasoning. "There is, apart from mere intellect,
   in the make-up of every superior human identity, a wondrous something that realizes without argument, frequently without what is called

education, an intuition of the absolute balance, in time and space, of the whole of this multifariousness, this revel of fools, and incredible make-believe and general unsettledness, we call the world. . . . [Of] such soul-sight and root-centre for the mind mere optimism explains only the surface" (304).

4. Ibid., 311.

5. Saver, J. L., and Rabin, J., "The Neural Substrates of Religious Experience," *Journal of Neuropsychiatry and Clinical Neurosciences,* 9 (1997): 498–510.

6. Alajouanine, F., "Dostoyevsky's Epilepsy," *Brain,* 86 (1963): 209–18.

7. James, 300.

8. Ibid., 302.

9. www.iands.org/nde.html.

10. www.nida.nih.gov/ResearchReports/Hallucinogens/halluc4.html. Jansen, K., "Using Ketamine to Induce the Near-Death Experience: Mechanism of Action and Therapeutic Potential," *Yearbook for Ethnomedicine and the Study of Consciousness,* 4 (1995): 55–81.

11. www.usdoj.gov/ndic/pubs/652/odd.htm#top.

12. LeDoux, J., *Synaptic Self* (New York: Viking, 2002), 210. Blakeslee, S., "Using Rats to Trace Anatomy of Fear, Biology of Emotion," *The New York Times,* November 5, 1996. Also available at www.cns.nyu.edu.

13. Damasio, A., *Descartes' Error* (New York: Avon Books, 1994), 118.

14. Phan, L., et al., "Functional Neuroimaging Studies of Human Emotions," *CNS Spectrums,* 9, no. 4 (2004): 258–66.

15. LeDoux, J., "Emotion, Memory and the Brain," *Scientific American,* 270 (1994): 34.

16. LeDoux, J., *The Emotional Brain* (New York: Simon & Schuster, 1996).

17. LeDoux, J., quoted in Daniel Goleman, *Emotional Intelligence* (New York: Bantam Books, 1996), 27. "Anatomically the emotional system can act independently of the neocortex. Some emotional reactions and emotional memories can be formed without any conscious, cognitive participation at all."

18. Bechara, A., et al., "Double Dissociation of Conditioning and Declarative Knowledge Relative to the Amygdala and Hippocampus in Humans," *Science*, 269 (1995): 1115–18.

19. Damasio, A., *The Feeling of What Happens: Body and Emotion in the Making of Consciousness* (New York: Harcourt Brace, 1999), 66.

20. Penfield, W., and Perot, P., "The Brain's Record of Auditory and Visual Experience," *Brain*, 86 (1963): 595–696. Bancaud, J., et al., "Anatomical Origin of Déjà Vu and Vivid 'Memories' in Human Temporal Lobe Epilepsy," *Brain*, 117 (1994): 71–90; Sengoku, A., Toichi, M., and Murai, T., "Dreamy States and Psychoses in Temporal Lobe Epilepsy: Mediating Role of Affect," *Psychiatry Clinical Neuroscience*, 51, no. 1 (1997): 23–26.

21. Ibid., Sengoku.

22. Note the similarity to Mr. C's complaint that his antique desk has been replaced by a cheap imitation.

Chapter 4: The Classification of Mental States

1. Damasio, A., *The Feeling of What Happens: Body and Emotion in the Making of Consciousness* (New York: Harcourt Brace, 1999), 340.

2. Ortony, A., and Turner, T. J., "What's Basic About Basic Emotions?" *Psychological Review*, 97 (1990): 315–31. Plutchik, R., "A General Psychoevolutionary Theory of Emotion," in Plutchik, R., and Kellerman, H., eds., *Emotion: Theory, Research, and Experience*, vol. 1, *Theories of Emotion* (New York: Academic, 1980): 3–33. Ekman, P., "Expression and Nature of Emotion," in *Approaches to Emotion*, Scherer, K., and Ekman, P., eds. (Hillsdale, N.J.: Erlbaum, 1984), 19–43.

3. Damasio, *The Feeling of What Happens*, 50.

4. Johnson-Laird, P., and Oatley, K., "Basic Emotions, Rationality, and Folk Theory," *Cognition and Emotion*, 6 (1992): 201–23.

5. Though purely mental constructs, they can be easily seen on fMRI, usually activating cortical regions also involved in the perception of external objects. If you look at a picture of George Washington and subsequently imagine the face of G.W., similar—but not necessarily identical—regions of visual cortex are activated.

6. Woody, E., and Szechtman, H., "The Sensation of Making Sense," *Psyche*, 8, no. 20 (October 2002). Also available at psyche.cs.monash.edu.au/v8/psyche-8-20-woody.html. An excellent discussion of the categorization of the feeling of knowing as a mental sensation.

7. books.guardian.co.uk/departments/biography/story/0,6000,674208,00 .html.

8. psyche.cs.monash.edu.au/v8/psyche-8-20-woody.html.

9. Hirstein, W., *Brain Fiction: Self-Deception and the Riddle of Confabulation* (Cambridge, Mass.: MIT Press, 2005), 97–99. Rasmussen, S. A., and Eisen, J. L., "The Epidemiology and Differential Diagnosis of Obsessive Compulsive Disorder," *Journal of Clinical Psychiatry*, 53 suppl. (1992): 4–10.

Chapter 5: Neural Networks

1. This is the conceptual location of the meaningless idea of the blank slate or tabula rasa. In the absence of experience, the equations may not contain specific values, but they do contain certain predilections or predispositions. Identical twins will have a greater degree of similarity of how their hidden layers process incoming information than strangers, but their individual perceptions will still be unique.

2. Moniz, E., "How I Came to Perform Leucotomy," *Psychosurgery* (1948): 11–14.

3. Olivecrona, H., the Nobel Prize in Physiology or Medicine 1949 presentation speech, www.nobel.se/medicine/laureates/1949/press.html.

4. Kopell, B., Reza, A., "The Continuing Evolution of Psychiatric Neurosurgery," *CNS Spectrum*, 5, no. 10 (2000): 20–31.

Chapter 6: Modularity and Emergence

1. An excellent summary of modularity and brain function is presented in Pinker, S., *How the Mind Works* (New York: Norton, 1997), and in Dennett, D., *Consciousness Explained* (Boston: Little, Brown, 1991). Pinker, 21: "The mind is organized into modules or mental organs, each with a specialized design that makes it an expert in one area of interaction with the world. The modules' basic logic is specified by our genetic program."

2. Zihl, J., and von Cramon, D., "Selective Disturbance of Movement Vision After Bilateral Brain Damage," *Brain*, 106 (1983): 313–40.

3. Johnson, S., *Emergence: The Connected Lives of Ants, Brains, Cities, and Software* (New York: Simon & Schuster, 2001). From the inside flap copy: "Emergence is what happens when an interconnected system of relatively simple elements self-organizes to form more intelligent, more adaptive higher-level behavior. It's a bottom-up model; rather than being engineered by a general or a master planner, emergence begins at the ground level. Systems that at first glance seem vastly different—ant colonies, human brains, cities, immune systems—all turn out to follow the rules of emergence. In each of these systems, agents residing on one scale start producing behavior that lays a scale above them: ants create colonies, urbanites create neighborhoods."

4. emergent.brynmawr.edu/eprg/?page=EmergenceReadingList is an excellent source for more detailed discussions of emergence.

5. Pinker, 30.

6. Cytowic, R., *Synesthesia: A Union of the Senses* (New York: Springer-Verlag, 1989), 1.

7. First quote: Lemley, B., "Do You See What They See?" *Discover*, 20 (1999): 12. Patricia Duffy's comment is from www.bluecatsandchartre usekittens.com.

8. Nabokov, V., *Speak, Memory* (New York: G. P. Putnam, 1966). Nabokov, 35: "The confessions of a synesthete must sound tedious and pretentious to those who are protected from such leakings and drafts by more solid walls than mine are. To my mother, though, this all seemed quite normal. The matter came up, one day in my seventh year, as I was using a heap of old alphabet blocks to build a tower. I casually remarked to her that their colors were all wrong. We discovered then that some of her letters had the same tint as mine and that, besides, she was often optically affected by musical notes. These evoked no chromatisms in me whatsoever."

9. Cytowic, R., an interview on www.abc.net.au/rn/talks/8.30/helthrpt/ hstories/hr080796.html.

10. publicaffairs.uth.tmc.edu/Media/newsreleases/nr2005/synesthesia.html .realmagick.com/articles/10/2210.html.

11. Ramachandran, V. S., and Hubbard, E., "Hearing Colors, Tasting Shapes," *Scientific American*, May 2003.

12. web.mit.edu/synesthesia/www/perspectives.html.

13. Duffy, P., *Blue Cats and Chartreuse Kittens: How Synesthetes Color Their Worlds* (New York: Henry Holt, 2001).

Chapter 7: When Does a Thought Begin?

1. Gammons, P., "The Science of Hitting," *The Boston Globe*, July 22, 2002. Also available at www.boston.com/sports/redsox/williams/july_22/The_ science_of_hitting+.shtml.

2. Will, G., *Men at Work* (New York: Harper Perennial, 1991), 193.

3. McLeod, P., "Reaction Time and High-Speed Ball Games," *Perception*, 16, no. 1 (1987): 49–59. "Laboratory measures of visual reaction time suggest that some aspects of high-speed ball games such as cricket are 'impossible' because there is insufficient time for the player to respond to unpredictable movements of the ball. Given the success with which some people perform these supposedly impossible acts, it has been assumed by some commentators that laboratory measures of reaction time are not applicable to skilled performers. An analysis of high-speed film of international cricketers batting on a specially prepared pitch which produced unpredictable movement of the ball is reported, and it is shown that, when batting, highly skilled professional cricketers show reaction times of around 200 ms, times similar to those found in traditional laboratory studies. Furthermore, professional cricketers take roughly as long as casual players to pick up ball flight information from films of bowlers. These two sets of results suggest that the dramatic contrast between the ability of skilled and unskilled sportsmen to act on the basis of visual information does not lie in differences in the speed of operation of the perceptual system. It lies in the organization of the motor system that uses the output of the perceptual system."

4. Mori, S., "Toward a Study of Sports Psychophysics." Presented at the 15th Annual Meeting of the International Society for Psychophysics, Tempe, Arizona, 1999. These minimum times are based upon physiological constraints and cannot be significantly shortened with additional training. The minor differences in reaction and swing time between novice and expert players do not explain the great differences in hitting ability.

5. Adair, R., *The Physics of Baseball* (New York: Harper Perennial, 2002), 42.

6. Nishida, S., and Johnston, A., "Influence of Motion Signals on the Perceived Position of Spatial Pattern," *Nature*, 397 (1999): 610–12.

7. Gomes, G., "The Timing of Conscious Experience: A Critical Review and Reinterpretation of Libet's Research," *Consciousness and Cognition*, 7 (1998): 559–95. Gomes, G., "Problems in the Timing of Conscious Experience," *Consciousness and Cognition*, 11 (2002): 191–97.

8. Jaeger, D., Gilman, S., and Aldridge, J., "Primate Basal Ganglia Activity in a Precued Reaching Task: Preparation for Movement," *Experimental Brain Research*, 95 (1993): 51–64.

9. Bahill, T., personal communication. See also Bahill, T., and Watts, R., *Keep Your Eye on the Ball: Curve Balls, Knuckleballs, and Fallacies of Baseball* (New York: W. H. Freeman, 2000).

10. "The average reaction times for initiating the swing and for changing the direction of the swing in response to the signals were, respectively, 206 and 269 milliseconds."

11. The transmission speed of nerve impulses may vary from 0.5 meters a second to more than 100 meters a second, depending on the type of nerve involved. For a two-meter-tall man, impulses from the great toe would reach the brain a minimum of ⅟₅₀ of a second later than from the nose (and perhaps considerably later), yet this difference is not noticed. The brain somehow bundles all the incoming information into a single *now* package.

12. To see a demonstration of color phi, go to www.yorku.ca/eye/colorphi .html.

13. Kolers, P., and von Grnau, M., "Shape and Color in Apparent Motion," *Vision Research*, 16 (1976): 329–35.

14. Damasio, A., "Remembering When," *Scientific American* (September 2002): 66.

15. At an average speed of only twenty-five miles per hour, a table tennis ball will travel the length of the table at the approximate limit of human reaction time. At world competition levels, the average speed of a hard-hit ball is around fifty-five miles per hour. www.jayandwanda.com/ tt/speed.html.

Chapter 8: Perceptual Thoughts: A Further Clarification

1. Budson, A. E., and Price, B., "Memory Dysfunction," *New England Journal of Medicine*, 352, no. 7 (2005). Semantic memories are now thought to be mediated by inferolateral temporal lobes, while episodic memories are mediated by the medial temporal lobes, anterior thalamic nucleus, mammillary body, fornix, and the prefrontal cortex.

2. Schacter, D., ed., *Memory Distortion: How Minds, Brains, and Societies Reconstruct the Past* (Cambridge, Mass.: Harvard University Press, 1997). Schacter, D., *Searching for Memory* (New York: Basic Books, 1996). Schacter, D., *The Seven Sins of Memory: How the Mind Forgets and Remembers* (Boston: Houghton Mifflin, 2001).

3. The difference between true thought and memory is evident during brain stimulation. Apply an electrode to the temporal lobe and subjects might recall old names, places, events, and faces. But stimulation does not reproduce the act of thinking. No one ever had forced reasoning or experienced a syllogism as the result of a seizure. The act of reasoning does not exist within the brain in the way that there are regions dedicated to processing sound and sight. It is a potential, an imbedded ability that emerges as an acquired skill, like the ability to read or play the accordion.

Chapter 9: The Pleasure of Your Thoughts

1. Olds, J., and Milner, P., "Positive Reinforcement Produced by Electrical Stimulation of Septal Area and Other Regions of Rat Brain," *Journal of Comparative and Physiological Psychology*, 47 (1954): 419–27.

2. Bozarth, M., "Pleasure Systems in the Brain," in Warburton, D. M., ed., *Pleasure: The Politics and the Reality* (New York: Wiley, 1994), 5–14. "Several neurotransmitters may be involved in the rewarding effects . . . , but dopamine appears to be the neurotransmitter essential for reward from activation of the medial forebrain bundle system. . . . The origin of the ventral tegmental dopamine system (i.e., ventral tegmentum) appears to provide an important neurochemical interface where exogenous opiates (e.g., heroin, morphine) and endogenous opioid peptides (e.g., endorphins, encephalin) can activate a brain mechanism involved in appetitive motivation and reward. This is not to suggest that all motivation effects of these rewards emanate from this single brain system, but rather this dopamine system represents one important mechanism

for the control of both normal and pathological behaviors." Nestler, E., and Malenka, R., "The Addicted Brain," ScientificAmerican.com, February 9, 2004.

3. Blood, A. J., and Zatorre, R. J., "Intensely Pleasurable Responses to Music Correlate with Activity in Brain Regions Implicated in Reward and Emotion," *Proceedings of the National Academy of Science*, 98, 20 (2001): 11818–23.

4. Goleman, D., "Brain Images of Addiction in Action Show Its Neural Basis," *The New York Times*, August 13, 1996.

5. Berridge, K., "Pleasures of the Brain," *Brain and Cognition*, 52 (2003): 106–28. An excellent review of the latest theories and neuroanatomy of brain reward systems.

6. Bechara, A., Damasio, H., and Damasio, A., "Emotion, Decision Making and the Orbitofrontal Cortex," *Cerebral Cortex*, 10, no. 3 (2000): 295–307.

7. Elliott, R., Dolan, R., and Frith, C., "Dissociable Functions in the Medial and Lateral Orbitofrontal Cortex: Evidence from Human Neuroimaging Studies," *Cerebral Cortex*, 10, no. 3 (2000): 308–17. "We suggest that the irrational sense of the rightness of a stimulus (which may relate to familiarity) is also associated with reward value . . . as with other regions of the prefrontal cortex, activity in the orbito-frontal cortex is most likely to be observed when there is insufficient information available to determine the appropriate course of action . . . that selection of stimuli on the basis of their familiarity and responses on the basis of a feeling of 'rightness' are also examples of selection on the basis of reward value."

8. Mantel, H., "Is the Particle There?" *London Review of Books*, July 7, 2005.

9. LeDoux, J., Romanski, L., and Xagoraris, A., "Indelibility of Subcortical Emotional Memories," *Journal of Cognitive Neuroscience*, 1 (1991): 238–43.

10. Kreek, M., Nielsen, D., Butelman, R., and LaForge, S., "Genetic Influences on Impulsivity, Risk Taking, Stress Responsivity and Vulnerability to Drug Abuse and Addiction," *Nature Neuroscience*, 8 (2005): 1450–57. "Chronic exposure to drugs of abuse causes persistent changes in the brain, including changes in expression of genes or their protein products, in protein-protein interactions, in neural networks, and in neurogenesis and synaptogenesis, all of which ultimately affect behavior. In rodents, there are inbred strains and selectively bred lines that readily

self-administer drugs of abuse (implying genetic vulnerability) as well as strains that do not readily self-administer drugs (implying genetic resistance). Different strains show differences in the cellular and molecular response to drugs. Genetic factors may also be involved in direct drug-induced effects, including alteration of pharmacodynamics (a drug's effects at a receptor, including the physiological consequences of receptor activity) or pharmacokinetics (a drug's absorption, distribution, metabolism and excretion) of a drug of abuse or of a treatment agent."

11. Kreek, 1450–57.

12. Brinn, D., "Israeli Researchers Discover Gene for Altruism," *Bulletin of Herzog Hospital* (January 23, 2005). Also available at www.herzoghospital.org.

Chapter 10: Genes and Thought

1. Bouchard, T., and McGue, M., "Genetic and Rearing Environmental Influences on Adult Personality: An Analysis of Adopted Twins Reared Apart," *Journal of Personality*, 58, no. 1 (March 1990). Bouchard, T., et al., "Sources of Human Psychological Differences: The Minnesota Study of Twins Reared Apart," *Science*, 250, no. 4978 (1990): 223–8.

2. Beckett, S., *The Unnamables* (New York: Grove Press, 1959), 418.

3. Drayna, D., "Is Our Behavior Written in Our Genes?" *New England Journal of Medicine*, 354, no. 1 (2006): 7–9. For a concise discussion of behavioral genetics, see www.ornl.gov/sci/techresources/Human_Genome /home.shtml.

4. Shumyatsky, G., et al., "Stathmin, a Gene Enriched in the Amygdala, Controls Both Learned and Innate Fear," *Cell*, 123 (2005): 697–709. "The knockout mice also exhibit decreased memory in amygdala-dependent fear conditioning and fail to recognize danger in innately aversive environments." www.nidcd.nih.gov/research/scientists/draynad.asp.

5. Carey, B., "Timid Mice Made Daring by Removing One Gene," *The New York Times*, November 18, 2005.

6. Zuckerman, M., and Kuhlman, D. M., "Personality and Risk-taking: Common Biosocial Factors," *Journal of Personality*, 68, no. 6 (2000): 999–1029. Perez de Castro, I., et al., "Genetic Association Study Between Pathological Gambling and a Functional DNA Polymorphism at

the D4 Receptor Gene," *Pharmacogenetics*, 7, no. 5 (1997): 345–48. "This work provides a new evidence of the implication of the dopaminergic reward pathways, now through the involvement of D4 dopamine receptor gene (DRD4) in the etiology of this impulsive disorder."

7. Zhang, L., Bao, S., and Merzenich, M. M., "Persistent and Specific Influences of Early Acoustic Environments on Primary Auditory Cortex," *Nature Neuroscience*, 4 (2001): 1123–30.

8. Chang, E. F., and Merzenich, M. M., "Environmental Noise Retards Auditory Cortical Development," *Science*, 300, no. 5618 (2003): 498–502.

9. www.hhmi.org/news/chang.html. "While the rat is not a perfect model of human auditory development, it does allow us to investigate the fundamental role of early sensory experience in mammalian auditory development. For example, we do know that exposing infant rats to specific sound stimuli can induce long-standing representational changes in the brain. Other researchers have shown that there are striking parallels in humans and other animals."

10. Dillon, S., "Literacy Falls for Graduates From College, Testing Finds," *The New York Times*, December 16, 2005. Also available at www.nytimes.com.

11. Merzenich has developed intriguing, albeit controversial techniques for presenting a new array of sounds to affected young children. It is his belief that an improperly developed auditory cortex can be reorganized to more efficiently and accurately process incoming speech. The key issues yet to be determined are the degree and duration of neural plasticity and ease of alteration of already established neural networks.

Chapter 11: Sensational Thoughts

1. Lakoff, G., and Johnson, M., *Philosophy in the Flesh* (New York: Basic Books, 1999), 4.

2. Blanke, O., et al., "Stimulating Illusory Own-Body Perceptions," *Nature*, 419 (2002): 269–70. Blanke, O., et al., "Linking Out-of-Body Experience and Self Processing to Mental Own-Body Imagery at the Temporoparietal Junction," *Journal of Neuroscience*, 25, no. 3 (2005): 550–57.

3. Awareness is the involuntary and selective perception of aspects of what the mind is doing at any instant. The difference between those

sensory inputs that do and do not reach consciousness isn't in the basic inputs, but in whether or not they are consciously felt.

4. Lakoff, 13. "Cognitive thought is the tip of an enormous iceberg. It is the rule of thumb among cognitive scientists that unconscious thought is 95 percent of all thought—and that may be a serious underestimate. Moreover, the 95 percent below the surface of conscious awareness shapes and structures all conscious thought. If the cognitive unconscious were not there doing this shaping, there could be no conscious thought."

5. The exact nature of these calculations isn't known, but synaptic transmission gives us some idea of the mathematical complexities involved. For any given type of neuron, some neurotransmitters are excitatory (encourage the cell to fire), while others are inhibitory (suppress the tendency of the neuron to fire). All compete with postsynaptic receptors, which also vary in their sensitivity and receptivity to the neurotransmitters. And so on. At every instant at every level of neural organization, this microscopic broth of pluses and minuses performs a vast number of constantly revised calculations.

Complex feelings are themselves the product of these calculations. In this light, it is easy to see the categories of *feeling of knowing* and *not knowing* as higher-level metaphoric equivalents of pluses and minuses. Familiarity, correctness, rightness, being on the right track, tip-of-the-tongue, and déjà vu are the pluses; the feelings of wrongness, strangeness, bizarreness, jamais vu, and "unrealness" are the minuses. The exact mix of these sensations will determine how we feel about an idea.

6. Pinker, S., *How the Mind Works* (New York: Norton, 1997), 70. "Sentences in a spoken language like English or Japanese are designed for vocal communication between impatient, intelligent social beings. They achieve brevity by leaving out any information that the listener can mentally fill in from the context. In contrast, the 'language of thought' in which knowledge is couched can leave nothing to the imagination, because it *is* the imagination. . . . So, the statements in a knowledge system are not sentences in English but rather inscriptions in a richer language of thought, 'mentalese.' "

Chapter 12: The Twin Pillars of Certainty: Reason and Objectivity

1. Goleman, D., *Emotional Intelligence* (New York: Bantam, 1997), 26.

2. O'Neil, J., "On Emotional Intelligence: A Conversation with Daniel Goleman," *Educational Leadership*, 54, no. 1 (September 1, 1996).

3. Goleman, 9.

4. Wilson, T., *Strangers to Ourselves: Discovering the Adaptive Unconscious* (Cambridge, Mass.: Harvard University Press, 2002), 16.

5. Ibid., 1.

6. Ibid., 16. "Making the 'unconscious conscious' may be no easier than viewing and understanding the assembly language controlling our word-processing computer program."

7. www.gladwell.com. Gladwell, M., *Blink* (New York: Little, Brown 2005), 256.

8. www.gladwell.com.

9. Gladwell, *Blink*, 14–16.

10. Ibid., 11–12.

11. www.edge.org.

12. Dijksterhuis, A., et al., "On Making the Right Choice: The Deliberation-Without-Attention Effect," *Science*, 311 (2006): 1005.

13. Anderson, L., "If You Really Think About It, Trust Your Gut for Decisions," *Chicago Tribune*, March 19, 2006.

14. Simons, D., and Chabris, C., "Gorillas in Our Midst," *Perception* (1999): 28.

15. Letter to Henry Fawcett, September 18, 1861, in Charles Darwin, *More Letters of Charles Darwin*, vol. 1, Darwin, F., and Seward, A., eds. (New York: D. Appleton, 1903), 194–96.

16. Gould, S., *The Lying Stones of Marrakech: Penultimate Reflections in Natural History* (New York: Harmony Books, 2000), 104–5.

17. Carey, B., "A Shocker: Partisan Thought Is Unconscious," *The New York Times*, January 24, 2006.

18. Ibid.

19. In writing this book, I have caught myself selecting facts to fit or support a preconceived idea that I've wanted to convey. This isn't a prudent admission if I want you to accept my ideas as being reasonable. On the other hand, it is an inescapable component of my thesis.

20. www.pbs.org/wgbh/pages/frontline/shows/altmed/interviews/weil.html.

21. Hultgren, L., www.pbs.org/wgbh/pages/frontline/shows/altmed/interviews /weil.html.

22. www.ions.org/publications/review/issue65/r65lora.pdf. Targ, R., and Katra, J., *Miracles of Mind: Exploring Nonlocal Consciousness and Spiritual Healing* (Novato, Calif.: New World Library, 1999), 193.

23. Gladwell, *Blink*, 16–17.

24. Frymoyer, J., "Back Pain and Sciatica," *New England Journal of Medicine*, 318, no. 5 (February 4, 1988).

25. Muller, R., "The Conservation Bomb," *Technology Review Online*, June 14, 2002. Also available at muller.lbl.gov/TRessays/05_Conservation_ Bomb.htm

26. americanradioworks.publicradio.org/features/climate/b6.html.

27. www.newscientist.com/hottopics/climate/climate.jsp?id–s99994888.

Chapter 13: Faith

1. I'm not referring to a specific intention such as in carrying an umbrella on a rainy day to avoid getting wet. In this context, purpose is synonymous with a conscious desire and intention not to get rained on. Carrying an umbrella can satisfy situational purposes, but rarely motivates you to jump out of bed in the morning to pursue your pet project.

2. Tolstoy, L., *My Confession My Religion* (Midland, Mich.: Avensblume Press, 1994).

3. Dawkins, R., from a debate with Archbishop of York, Dr. John Habgood, *The Nullifidian* (December 1994). Also available at www.world-of-daw kins.com/religion.html.

4. Dawkins, R., quoted in *The Guardian*, October 3, 1998.

5. en.wikiquote.org/wiki/Stephen_Hawking.

6. Davies, P., "Universal Truths," *The Guardian*, January 23, 2003.

7. Davies, P., *God and the New Physics* (London: Penguin, 1990), 189.

8. Weinberg, S., *The First Three Minutes* (New York: Basic Books, 1993).

9. Kass, L., *Toward a More Natural Science* (New York: Free Press, 1988).

10. www.pbs.org/wgbh/questionofgod/voices/index.html.

11. Lewontin, R., "Billions and Billions of Demons," *The New York Review of Books*, January 9, 1997. Also available at www.nybooks.com/articles/1297.

12. Quoted in Brown, A., *The Guardian*, April 17, 2004, in a review of Dennett's writings. books.guardian.co.uk/review/story/0,12084,1192975,00.html.

13. I don't personally see the point of Weinberg's comment that our increasing understanding of the world makes it seem all the more pointless. I understand his arguments, but rather than evoking a sense of personal despair, they make me laugh at the ridiculousness of believing that we can understand why we are here. If there is some meaning or purpose, please don't tell me. If a sign were to drop out of the sky and tell me what the meaning of life was, and it wasn't to my liking, I'd be much more disappointed than if I didn't see the sign. Not knowing gives me license to pursue the ridiculous. My basic personality has prompted my writing a book pointing out that the determination of pointless or purposeful cannot be a purely rational decision. I enjoy the basketball-gorilla video because it confirms my deepest suspicions that we are more likely to see what we want to see and less likely to see that which isn't of interest—including purpose or pointlessness. My differences with Lewontin might seem like reasoned arguments, but they are ultimately reflections of unavoidable different ways of seeing the world.

14. Slack, G., an interview with Dennett, *Salon.com*, February 8, 2006. www.salon.com/books/int/2006/02/08/dennett.

15. Barlow, N., ed., *Charles Darwin* (New York: Norton, 1993). www.update.uu.se/~fbendz/library/cd_relig.html.

16. Gould, S., "The Validation of Continental Drift," in *Ever Since Darwin: Reflections in Natural History* (London: Penguin, 1991), 161. "During the period of nearly universal rejection, direct evidence for continental drift—that is, the data gathered from rocks exposed on our continents—was every bit as good as it is today. . . . In the absence of a plausible mechanism, the idea of continental drift was rejected as absurd. The data that seemed to support it could always be explained away. . . . The old data from continental rocks, once soundly rejected, have been exhumed

and exalted as conclusive proof of drift. In short, we now accept continental drift because it is the expectation of a new orthodoxy. I regard
this tale as typical of scientific progress. New facts, collected in old ways
under the guidance of old theories, rarely lead to any substantial revi-
sion of thought. *Facts do not 'speak for themselves,' they are read in the
light of theory.*" (Italics mine.)

17. Lovgren, S., "Evolution and Religion Can Coexist, Scientists Say," *National Geographic News*, October 18, 2004. Also available at news.nation
algeographic.com/news/2004/10/1018_041018_science_religion.html.

Chapter 14: Mind Speculations

1. Rundle, B., *Why There Is Something Rather Than Nothing* (Oxford: Oxford University Press, 2004). Bebe, an Oxford University philosophy
professor, argues that the question, and the attempts to answer it, consistently take language beyond the bounds of meaningfulness; detaching
familiar words from their usual roles so that they no longer express in-
telligible possibilities. Calling a space surrounding an object nothing
strips *nothing* of any real meaning.

2. www.pbs.org/wgbh/nova/origins/universe.html.

3. Gott, J., et al., "Will the Universe Expand Forever?" *Scientific American*
(March 1976), 65.

4. Hawking, S., *A Brief History of Time* (New York: Bantam Books, 1988),
140–41.

5. www.kirjasto.sci.fi/ikant.html.

6. Bisson, T., "They're Made Out of Meat," a short story in *Omni* (April
1991). Also available at www.terrybisson.com/meat.html.

7. Nearly three hundred years ago, philosopher and mathematician Gott-
fried Leibnitz hypothesized a machine capable of conscious experiences
and perceptions. He said that even if this machine were as big as a mill
and we could explore inside, we would find "nothing but pieces which
push one against the other and never anything to account for a perception." Our present-day conceptualization of emergence isn't much better. Perhaps if we do develop a better idea of how emergence is
physically manifest we will come up with a more meaningful classification of mental states. Until then, we are better off sticking with the

clumsy notion of "real" but subjective, rather than trying to squeeze as yet ill-understood phenomena into equally confusing and misleading physical and mental categories.

8. Searle, J., *Mind: A Brief Introduction* (Oxford: Oxford University Press, 2004), provides an excellent overview of the various mind-body arguments that have arisen in an attempt to explain why the mind is or isn't more than the brain that created it.

9. Libet, B., *Mind Time* (Cambridge, Mass.: Harvard University Press, 2004).

10. Any experiment on free will reflects the inherent problem it is trying to study. The most obvious paradox is whether or not we are free to choose which experiments best demonstrate the presence or absence of free will. In order to properly design a study, we have to believe that we have the freedom to choose good from bad evidence. If we don't believe in free will, we must concede that any experiment we choose will be beyond our control—negating the principles of rationality and objectivity crucial to scientific method. The middle ground—an entirely neutral view of free will—goes against our understanding of motivation and why we would undertake the study in the first place. All studies on free will are handicapped at the start.

11. Libet, 109.

12. www.tourettes-disorder.com/symptoms/coprolalia.html.

13. www.tourettes-disorder.com/blogs/2005/03/regarding-coprolalia-and-use -of.html.

Chapter 15: Final Thoughts

1. www.edge.org/q2006/q06_12.html.

2. Overbye, D., "From a Physicist and New Nobel Winner, Some Food for Thought," *The New York Times*, October 19, 2004. Also available at www .nytimes.com/2004/10/19/science/19phys.html.

Acknowledgments

It is impossible to trace the origins of a book that has percolated for so many years. There are, needless to say, many people who inspired and helped me with this project whom I would like to thank. They include my colleagues at the San Francisco Philosophical Society as well as Jonathon Keats, Kevin Berger, Peter Robinson, David Steinsaltz, Richard Segal, and Herbert Gold.

I am extremely fortunate to have Jeff Kellogg as my literary agent; he has been a source of constant encouragement and instrumental in converting scribbles in a personal journal into the book's present structure. Nichole Argyres, my editor, and her assistant, Kylah McNeill, have provided enthusiastic support and have greatly improved my original manuscript.

Unfortunately, I cannot directly thank the many patients who have prompted me to ask the questions at the heart of this book. For those patients who might be reading it, please know that I am forever indebted.

Above all, I express my deepest thanks to my wife, Adrianne, who has been my continuing inspiration, staunchest supporter, and level-headed critic. It is impossible for me to adequately express the depth of my gratitude and appreciation. So, thanks, Adrianne.

Index